MICROBES

The Life-Changing Story of Germs

Phillip K. Peterson

Prometheus Books
Guilford, Connecticut

(PB) Prometheus Books

An imprint of The Rowman & Littlefield Publishing Group, Inc.
4501 Forbes Boulevard, Suite 200
Lanham, Maryland 20706
www.rowman.com

Distributed by NATIONAL BOOK NETWORK

British Library Cataloguing in Publication Information Available

Library of Congress Cataloging-in-Publication Data

Name: Peterson, Phillip K., author.
Title: Microbes : the life-changing story of germs / Phillip K. Peterson.
Description: Lanham, MD : Prometheus, [2020] | Includes bibliographical references and index. | Summary: "With straightforward and engaging writing, infectious diseases physician Phillip Peterson surveys how our understanding of viruses has changed throughout history, from early plagues and pandemics to more recent outbreaks like HIV/AIDS, Ebola, and the Zika virus. Microbes also takes on contemporary issues like the importance of vaccinations in the face of the growing anti-vaxxer movement, as well as the rise of cutting-edge health treatments like fecal transplants. Microbes explains for general readers where these germs came from, what they do to and for us, and what can be done to stop the bad actors and foster the benefactors"—Provided by publisher.
Identifiers: LCCN 2019057113 (print) | LCCN 2019057114 (ebook) | ISBN 9781633886346 (cloth) | ISBN 9781633886353 (epub)
Subjects: LCSH: Microorganisms—Popular works. | Medical microbiology—Popular works. | Vaccines—Popular works.
Classification: LCC QR56 .P48 2020 (print) | LCC QR56 (ebook) | DDC 616.9/041—dc23
LC record available at https://lccn.loc.gov/2019057113
LC ebook record available at https://lccn.loc.gov/2019057114

To Anders, Sonja, Ilsa, and Svea

"It is the microbes who will have the last word."
—Louis Pasteur

"If microbial life were to disappear, that would be it—instant death for the planet."
—Carl Woese

CONTENTS

FOREWORD

Do a search of the major online bookstores and you'll find more than ten thousand titles addressing one or more aspects of microbes, including those related to infectious diseases. Imagine trying to read one of these books each week; it would take you at least 192 years to complete that task. So I present a major time-saver: *Microbes: The Life-Changing Story of Germs*, which summarizes all the important information you need to know in an easy-to-read and understandable book. It features a fun narrative that highlights the content and message of all of these books about microbes. It is a microbial history lesson, a personal journal, a worldwide travelogue, a science textbook, and an entertainment guide, all wrapped into one. Whether you are an infectious disease expert, a general practice physician or nurse, a microbiologist, a teacher or student, or even a member of the general public with an interest in the fascinating, dynamic world of microbes and human, animal, and environmental health, this book is a must read! You will take delight in what you learn about your own health and how the ever-evolving microbial world affects it.

Dr. Peterson is a storyteller who also has an in-depth understanding of "the good, the bad, and the ugly" of microbes. This book is conveniently divided into three parts. In the first part, "Intimate Friends," we learn how microbes are responsible for our current oxygen-rich environment that supports all life and how they are our intimate bodyguards and critical companions in making our very existence possible. In the second part, "Mortal Enemies," we learn how microbes can kill us, as well as all other living species on earth. In the last part, "Germs in the Future," we are treated to a futuristic story of how we can harness the power of microbes to make us healthier and even safer, while also recognizing that germs are still in the driver's seat with regard to the ever-dynamic world of microbial evolution.

We are treated to many fascinating facts in this book, but rather than challenging us to be scientists or healthcare providers, Dr. Peterson entertains and informs in a way that makes it hard to stop turning the page. Many readers likely will be surprised to learn of the essential role microbes have played in human history. This includes how microbes keep us healthy—yes, most are our dear friends and even our protectors, not killers or health thugs—and in some cases even help us become healthier. Most microbes are beneficial to human, animal, and plant health and planetary health overall. Today we are bombarded with advertising for a variety of disinfectants that want us to believe the only good microbe is a dead microbe. This book is an authoritative guide to understanding and appreciating how microbes benefit us. Imagine how your life, or that of a loved one or colleague, might one day be saved by a friendly "therapeutic virus," called a phage, that attacks the antibiotic-resistant infection caused by bacteria for which we no longer have an effective antibiotic.

But make no mistake, there are microbes that are our enemies. In fact, today there are more than 1,400 recognized infectious diseases—as well as others yet to be discovered or for which the microbes have yet to evolve into disease causers. Yes, these disease-causing microbes are but a small minority of the total number of microbes in our world, but their damaging impact can't be denied or minimized. We learn in narrative detail the deathly impact of microbial-caused diseases like smallpox, plague, Ebola, HIV/AIDS, tuberculosis, malaria, influenza, cholera, Zika, dengue, and antimicrobial-resistant infections. The viruses, bacteria, and parasites that cause these diseases are as dangerous and capable of causing serious and even societal-disruptive impacts on humans, animals, and plants as any major weapon system our military possesses today. A devastating influenza pandemic—a worldwide epidemic—could result in more global deaths in just a few months than the detonation of a nuclear warhead somewhere in the world. Dr. Peterson provides us with a clear and compelling story of why the future of infectious diseases is seriously challenging, even with our modern medical research and technology accomplishments.

This book will help you conclude that we must deeply respect microbes and understand in much more detail how they help us and how they kill us. The importance of vaccines and antibiotics, and why we must continue to invest in their development, is described in the clearest of terms. Just think of the incredible impact that vaccines against infectious diseases have had; the Centers for Disease Control and Prevention estimates that vaccines given to infants and young children over the past two decades will prevent 322 million illnesses, 21 million hospitalizations, and 732,000 deaths over the course of those people's lifetimes. Compelling? I think so. Nonetheless, today we have severe and vocal critics of vaccines, who through deceptive messages not based on any reliable scientific studies strongly discourage the use of vaccines. Dr. Peterson addresses this issue head-on and provides us with the information we need to challenge these dangerous antivaccine voices.

If you have one book to read to understand your health and the world around you, *Microbes: The Life-Changing Story of Germs* should be at the top of your list. It is a gift to all of us who live in a world of microbes!

—Michael T. Osterholm, PhD, MPH, Regents Professor, McKnight Endowed Presidential Chair in Public Health, director of the Center for Infectious Disease Research and Policy, distinguished university teaching professor of environmental health sciences for the School of Public Health, professor of technological leadership for the Institute College of Science and Engineering, and adjunct professor of the medical school at the University of Minnesota

NOTE TO THE READER

This book provides discussions regarding various illnesses and is not intended to be used to diagnose or treat any condition. If you have any symptoms or concerns about an illness, please consult a qualified health professional.

ACKNOWLEDGMENTS

A few years ago, I gave microscopes to my four grandchildren (to whom this book is dedicated) for Christmas. Watching their excitement as the microscopic world opened to them, as it first thrilled me long ago, was an important inspiration to write this book.

During the course of the writing, as I shared "germ gems" with many friends and colleagues, I realized that almost everyone is intrigued by microbes. (I say "almost" because I wasn't convinced that my best friend—my wife, Karin—appreciated hearing germ gems every time we went for a walk during the past several years. But, good-naturedly, she played a key role in completing the book, and she remains my best friend.)

I also want to thank a major contributor to the book, Scott Edelstein, my writing consultant and literary agent, whose advice on how to make stories about germs come alive was invaluable. Without his help and wonderful sense of humor, *Microbes: The Life-Changing Story of Germs* wouldn't have come to light. I'm also grateful for the input and cheerful guidance of my editors, Jacob Bonar, Kellie Hagan, and Andrew White.

Finally, I want to thank three colleagues and cherished friends of more than four decades. Mike Osterholm, one of the world's top authorities in the field of infectious diseases epidemiology, provided the foreword. Not only is it an important contribution to the book, but it reflects his generosity. Paul Quie, a giant in the field of infectious diseases, has been my mentor since day one of my academic career. He and David Williams— the most gifted teacher, internist, and infectious disease specialist I know—provided many insights. These insights reflect not only the topics in *Microbes: The Life-Changing Story of Germs* but, over the years, the much bigger subject of human kindness.

INTRODUCTION

How Tiny Creatures Make a Huge Impact

"Because we humans are big and clever enough to produce and utilize antibiotics and disinfectants, it is easy to convince ourselves that we have banished bacteria to the fringes of existence. Don't you believe it. Bacteria may not build cities or have interesting social lives, but they will be here when the Sun explodes. This is their planet, and we are on it only because they allow us to be."—Bill Bryson

"One side of a story is hardly a story at all. It's more like propaganda when you think about it."—Daniel L. Robinson

This book began germinating three decades ago in my daughter's fourth-grade classroom. She had arranged for me to give a talk to her class on what I did for a living. I was then a newly minted infectious disease specialist. I knew that her classmates had heard how dangerous germs can be—so to help them get a broader understanding, I gave a talk called "Germs Are Your Friends."

I brought along a replica of the microscope invented by Antonie van Leeuwenhoek. (The microscope itself is surprisingly small—only about three inches tall.) I also brought a couple of dozen petri dishes for the kids to cough or spit on. After two days in an incubator, they would demonstrate the wide variety of bacteria and fungi the kids harbored in their mouths.

My talk was well received by the students, but their teacher became visibly uneasy when I explained that most germs are totally harmless, and many are even good for you. Thus she ushered me out of the room before I could answer all the kids' questions. [1]

If I were to give a similar talk today, I wouldn't call it "Germs Are Your Friends." I'd call it "How Germs Have Already Saved Your Life," because germs protect us from illness and death every day of our lives. Recent advances in the fields of molecular biology, evolutionary biology, and ecology have led to a new appreciation of the pivotal role that microbes play in human health and well-being.

Beginning in the last quarter of the twentieth century, extraordinary scientific breakthroughs revealed the crucial role germs play in our health and the health of our planet. But at the same time, an acceleration of new and frightening infections has erupted throughout the world. This book tells the thrilling but sometimes terrifying stories of these germs—both the good and the bad actors.

The life-changing power of germs—microscopic creatures also known as *microbes* or *microorganisms*—is difficult to overstate. Indeed life itself had its genesis with germs. But at the same time that they continue to promote health, microbes threaten the lives of virtually all other life-forms.

Charles Sydney Burwell, a former dean of Harvard Medical School, is recognized for these words of wisdom: "My students are dismayed when I say to them, 'Half of what you are taught as medical students will in ten years have been shown to be wrong. And the trouble is, none of your teachers know which half.'"

I graduated from Columbia University College of Physicians and Surgeons more than four decades ago. Back then, we were taught that germs caused all kinds of dangerous infectious diseases, many of which can be fatal.

Today, that observation—that germs are our mortal enemies—hasn't changed. But it turns out that this is only a small part of the story. Because of scientific advances in just the past few years, we now know that the vast majority of germs (bacteria, archaea, viruses, fungi, and protists) are either harmless or genuinely essential to human health. They are our intimate friends.

And they are everywhere—and have been since life began. Bacteria inhabit every surface of your body. If you're healthy, your gastrointestinal tract alone is inhabited by about *forty trillion* bacteria. That's about the same number of cells in your entire body. Virtually all of these bacteria are either harmless or beneficial to your health.

Germs are our ancient common ancestors. They outweigh all animals on Earth combined. And they often outsmart us, as we've seen with the recent spread of antibiotic-resistant microbes.

Germs are also vital to our planet. They are critical players in the health of all ecosystems—from our bodies to our oceans.

A common misconception at the time I entered medical school in 1966 was that infectious diseases had been conquered. One of the most famous quotes in modern medicine, attributed at the time to the then surgeon general, William H. Stewart, reflects that mistaken notion: "It's time to close the books on infectious diseases, declare the war against pestilence won, and shift national resources to such chronic problems as cancer and heart disease." The fact is, however, that William Stewart never said this. And nothing could have been further from the truth.

In 1992, a publication from the Institute of Medicine, *Emerging Infections: Microbial Threats to Health in the United States*,[2] set the record straight. The report, intended as a wake-up call for the U.S. Congress, detailed the alarming increase in infectious diseases during the preceding twenty-five years.

As you will discover in this book, germs have played—and continue to play—an enormous role in human history. Smallpox alone killed more people than all the wars *throughout history* combined. (Yet the virus that caused smallpox is the one and only virus that has actually been conquered, thanks to vaccination.) And the bacterium that caused the Black Death, *Yersinia pestis*, killed twenty-five million Europeans in the mid-fourteenth century. (The current population of all of New England is about fifteen million.)

In the years since I finished my infectious diseases training in 1977, the diseases that have emerged include *Clostridioides difficile* infection, Legionnaires' disease, Lyme disease, HIV/AIDS, Ebola virus disease, West Nile virus encephalitis, SARS, methicillin-resistant *Staphylococcus aureus* (MRSA), and Zika virus infection, to name just a few.

But humans' number one mortal enemy, both in the past and right now, is influenza virus, which my colleague Mike Osterholm calls the "king of infectious diseases." The 1918–1919 flu epidemic killed more people than World War I. The next big flu epidemic is literally waiting in the wings, since it will be initially spread by birds.

While this book tells the story of how human life—and human history—have been profoundly altered by our microbial mortal enemies, it also tells the inspiring and often-surprising story of our intimate friends. These are the germs that keep us healthy; that can help us become healthier; and, in some cases, that made human life possible in the first place.

As this newly understood story reveals, the great majority of germs are beneficial to human and planetary health. Germs offer us hope for new vaccines—and better and longer lives. They may also be part of the solution to climate change. This book looks at a variety of microbes that can transform human life for the better, and the technology currently being developed will encourage this transformation.

* * *

I've spent almost four decades as an infectious disease specialist on the front lines, battling germs. My forty years as an infectious disease consultant and a researcher at the University of Minnesota spanned the era of most emerging infections. (You will read about some of them in the "Mortal Enemies" section of this book.) I also had the good fortune to observe the birth of the era of the human microbiome, a topic discussed in the "Intimate Friends" section. Throughout these years, I've become increasingly amazed by how cunning microbes can be in doing us harm. But I've also become increasingly aware of how critically important many germs are to human lives, our food supply, and the health and survival of the entire planet.

This book tells the awesome story of germs—a story of benefactors and adversaries. It looks both to the past and into the near future. It is a factual story, but one riddled with mystery. As you'll discover, new (and currently unknown) infections will continue to emerge that will impact the lives of humans, animals, and plants. But, as you'll also read, the tale of germs is also a story of hope—about saving your life, our species, and planet Earth.

Part One

Intimate Friends

I

THE TREE OF LIFE

To see a World in a Grain of Sand
And a Heaven in a Wild Flower,
Hold Infinity in the palm of your hand
And Eternity in an hour.
—William Blake

Astronomers tell us that there are about as many stars in the universe as there are grains of sand on Earth. Both are estimated at about 10^{24}—one followed by twenty-four zeros. That is a very, very large number.

But scientists have used electron microscopes to scan individual grains of sand and have discovered that each grain harbors thousands of bacteria. By some estimates, the number of bacteria that inhabit our planet equals 10^{30}—that's one followed by thirty zeros, or one nonillion. That's roughly a million times the number of all the stars in the universe.

But, first, what exactly is a germ?

IN THE BEGINNING

"He who can no longer pause to wonder and stand in awe, is as good as dead; his eyes are closed."—Albert Einstein

While the terms *germ, microorganism,* and *microbe* are synonyms and are used interchangeably, *germ* is particularly fitting. The word comes from the Latin word *germen,* which refers to the sprout of a plant. This is the same Latin word that gave us *germinate.* As you will read, germs gave rise not only to plants but to all living things, including animals like us. In this book, the term *germ* refers to creatures that are microscopic, that is, out of sight of the unaided human eye.

The word *germ* first appeared in English in the seventeenth century. Back then, germs were considered good and positive things, as in "the germ of an idea."

But in the nineteenth century, the germ theory of disease took hold, and from then on germs have had a bad name. That's a shame, because many germs are good for us, and relatively few can do us harm. As we will see, however, all types of germs deserve our healthy respect.

THE ORIGIN OF LIFE AND OF SPECIES

The smallest unit of life is called an *organism*. An organism is composed of one or more cells, and all organisms do three basic things: metabolize (break down molecules to obtain energy), respond to stimuli in their environment, and reproduce. Or, to oversimplify things a bit, they all eat, fight or flee (or both), and give birth. Their long-term survival also requires them to adapt to their environment by evolving.

The question *How was the first living organism created?* has yet to be answered—though we do know that it occurred about 3.8 billion years ago.

Charles Darwin, the father of evolutionary biology, puzzled over this question and considered it the most fundamental of all questions, but was unable to answer it. Later researchers carried out all kinds of elegant experiments, concocting chemical mixtures resembling what is often called *primordial soup*, hoping to generate life. But so far, all have failed. The origins of life remain a mystery.

Darwin's theory of evolution was based largely on his astute observations of the features of living organisms and of fossils—previously living creatures encased in rock, amber, or other material. But for Darwin, the fossil record stopped at rocks that corresponded to the end of the Precambrian Period, about 570 million years ago. In contrast, modern techniques in geology, paleontology, and molecular biology have revealed that the oldest fossil in the world is over *six times* that old—3.7 billion years old. That organism, recently found in stromatolites in Greenland, was a germ called cyanobacteria. (In 2017, Canadian researchers described putative

fossilized microorganisms in sedimentary rocks in Quebec that they estimate are 3.8 to 4.3 billion years old, a claim that is disputed by other
scientists.)

MICROORGANISMS AND THE BIG PICTURE

"What you see is that the most outstanding feature of life's history is a
constant domination by bacteria."—Stephen Jay Gould

Microbes are at the very root of what is called the Tree of Life. As you
can see in the illustration on the following page, living organisms are
classified into three major groups called domains: Bacteria, Archaea (derived from the Greek word meaning "ancient"), and Eukarya. Both bacteria and archaeans are extremely tiny; each organism is no larger than a
single cell—and, thus, invisible to the unaided eye. Each of these creatures also has only one chromosome—which along with its DNA resides
in what is called cytoplasm.

In contrast, most eukaryotes are multicellular—though there are sin-
gle-celled ones, such as some types of protists. (It should be noted that
Fungi—which, along with Animalia and Plantae, sit at the top of the
Eukarya limb of the Tree of Life—aren't all microscopic. In fact, the
largest organism on Earth is a fungus named *Armillaria ostoyae* that
occupies more than two thousand acres of forest floor in Eastern Oregon.
Other macroscopic fungi that you are more familiar with are mushrooms.
Edible mushrooms, of course, have nutritional value. And studies at
Johns Hopkins University have found that psilocybin, the active ingredi-
ent of "magic mushrooms"—or "shrooms"—relieves depression and anxiety.[1])

Humans, animals, plants, fungi, and protists are all types of eukaryotes. The distinguishing feature of all eukaryotes—including human beings—is that each of their cells has a nucleus that houses multiple
chromosomes and most of their DNA. (Human beings have forty-six
chromosomes; dogs have seventy-eight; cats and pigs have thirty-eight.)

But single-celled microscopic organisms called germs appear in all
three domains of the Tree of Life, and we now know that some of them
(bacteria and archaea) gave rise to the eukaryotes. Recent studies suggest
that about 3.8 billion years ago, a microbe called LUCA (last universal

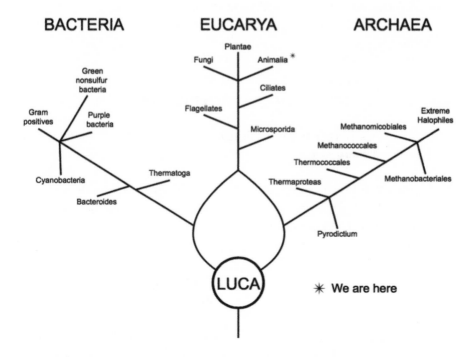

BACTERIA   EUCARYA   ARCHAEA

Simplified version of the phylogenetic Tree of Life originally proposed in the 1970s by Carl Woese and his colleagues. At the base of the tree is the last universal common ancestor (LUCA), postulated in 2016 by a team of researchers led by William Martin. (Using new methods to generate genome sequences and a super-computer, Jillian Banfield and her colleagues recently proposed a new view of the tree of life that includes ninety-two Bacteria phyla, twenty-six Archaea phyla, and five Eucarya supergroups.)

common ancestor)—sometimes referred to as the microbial Eve—emerged. And around two billion years after that, a merger of bacteria and archaea occurred, giving the Eucarya their start in life. Therefore, microbes had the planet all to themselves for about two billion years.

In his book *The Tangled Tree: A Radical New History of Life*, David Quammen provides a fascinating account of the breakthrough discovery of Archaea by Carl Woese, as well as the highly controversial proposal by Lynn Margulis that the energy-generating organelles in prokaryotic cells called mitochondria are derived from another sort of bacterium.[2] Quammen also makes a solid case for the importance of the transfer of genes between microbes as a driving force of evolution.

So we now know that, without germs (bacteria and archaea), we human beings wouldn't be here.

THE OUTLIERS OF LIFE

Technically, there is also a fourth type of germ—the virus. Viruses are serious outliers. In fact, most biologists don't even consider viruses to be living organisms—which is why they don't appear in the Tree of Life.

Unlike living creatures—even minuscule, one-celled ones—viruses don't possess metabolic machinery of their own. Nor do they have the capacity to reproduce on their own. Instead, they commandeer the host cells that they infect. In the words of virologists Marc H. V. van Regenmortel and Brian W. J. Mahy, viruses lead "a kind of borrowed life."

Like other germs, viruses are very simple and very tiny—too small to be seen by the eye, or even under a standard microscope. They consist of nothing more than some genes packaged inside a protein coat. They have an evolutionary history of their own, dating back as far as the origin of cellular life. In fact, viruses, bacteria, and archaea have been coevolving ever since, and about 1.5 billion years ago their evolutionary path was joined by the eukaryotes.

There is a mind-blowing number of types or species of viruses—by one estimate, hundreds of millions, by another at least a billion.[3] Only about 5,000 of them have been studied and described in detail so far.

A small number of viruses are deadly to humans, many are innocuous, and, as with other types of germs, some are highly beneficial. In fact, about 8 percent of the human genome consists of endogenous retroviruses that inserted themselves into our DNA eons ago. And some of this viral DNA is essential for physiological functions that are necessary for our survival.

The discovery in 2003 of giant viruses, called mimiviruses, with hundreds or even thousands of genes, shattered the existing definition of living organisms, paving the way for placing viruses in the Tree of Life. Moreover, recent studies by Gustavo Caetano-Anolles and his colleagues at the University of Illinois suggest that viruses and bacteria both descended from an ancient cellular life-form.[4]

Mimiviruses are big enough to be seen under a regular microscope. And some are even larger than bacteria. In 2017, Frederick Schulz and his colleagues described mimiviruses called Klosneuviruses (they were re-covered from sludge in the eastern Austria town of Klosterneuberg) with genomes resembling those of members of the Tree of Life.[5] Arguments are ongoing whether viruses belong as a fourth domain in the tree. But like the vast majority of smaller viruses, none of the giant viruses are known to make humans sick.

2

IT'S A MICROBIAL WORLD

"Where the telescope ends, the microscope begins. Which of the two has the grander view?"—Victor Hugo

In 1850, the French scientist Louis Pasteur was the first researcher to hypothesize that germs cause disease and to carry out experiments to test that hypothesis. But the experiment that definitively proved the germ theory of disease was carried out in 1875 by a thirty-three-year-old German country doctor, Robert Koch.

At the time, anthrax was decimating cattle, as well as causing serious trouble in humans.[1] Koch isolated a bacterium, which he named *Bacillus anthracis*, from the blood of a dead animal and grew it in a pure culture—a scientific breakthrough in itself. He then inoculated a healthy rabbit with the bacterium. The rabbit developed anthrax, after which Koch took blood from the rabbit and found the bacterium in the blood. Thus, Koch determined that the bacterium was the cause of the disease. (The process of isolating a microbe from a dead animal, using it to transmit the disease to a healthy animal, and then isolating it again became known as "Koch's postulates.")

In 1882, Koch went on to discover the bacterium that causes tuberculosis (*Mycobacterium tuberculosis*). And in 1897 he explained the cause of the dreaded bubonic plagues of the Middle Ages, involving transmission of the plague bacillus, *Yersinia pestis*, via lice from infected rats.

As a result of the research of Pasteur and Koch, the field of microbiology took off. Physicians and scientists discovered the germs that caused many infectious diseases—and devised methods to prevent or treat them.

The first Nobel Prize in Physiology or Medicine was awarded in 1901 to Emil von Behring for his discovery of the bacterial toxin that causes diphtheria. In the Nobel's first twenty years, bacteriology took half the prizes—and from 1921 to 1940 almost as many.

However, research into the smallest germs, viruses, lagged behind that of larger microbes. This was largely because viruses can't be seen with a conventional microscope, one that uses light.

The possibility that such really tiny infectious agents existed was first suggested by a Russian biologist, Dimitri Ivanovsky. He determined in 1892 that a disease of tobacco plants was caused by an agent so small that it passed through filters that captured larger microorganisms, such as bacteria. In 1898, Martinus Beijernick, a Dutch microbiologist, coined the term *virus* for such unfilterable infectious agents.

In 1931, the electron microscope, which uses a beam of electrons instead of light, was invented. (Electron microscopes can magnify up to 10,000,000 times, whereas most light microscopes can magnify no more than 2,000 times.) And in 1939, the first virus—tobacco mosaic virus— was seen using this instrument.

Surprisingly, the number of types of microbes that cause disease, called pathogens, is miniscule. Of the estimated tens of millions of bacterial species, for example, only about 1,400 cause illness in humans. As for the many millions of species of one-celled archaea, only one so far has been identified as a cause of human infection. And while viruses receive much negative press, most of them aren't our enemies. One group of viruses, called *bacteriophages*, is enormously helpful to human beings. Bacteriophages wreak havoc on bacterial pathogens, and thus are enemies of our enemies.

Bacteriophages are particularly abundant in seawater, where they far outnumber every other biological entity. For example, a liter of seawater collected from marine surface waters typically contains at least ten billion bacteria and one hundred billion viruses—the vast majority of which remain uncharacterized. All but overlooked until this century, these bacteriophages now are considered drivers of global biogeochemical cycles of carbon, nitrogen, sulfur, and oxygen. Along with unicellular eukaryotes—plankton and algae—they play an enormous role in shaping the earth's atmosphere and in sustaining marine food webs. Bacteriophages are also indirectly responsible for limiting global warming. They reduce the amount of carbon dioxide in the atmosphere by about three billion tons per year.

Bacteria are about as small as you can imagine. Between one thousand and one hundred thousand can fit on an average pinhead. But depending on which virus you're talking about, a million or more virus particles can fit on the head of a pin. About five million bacteria live in a teaspoon of salt water. (By the way, without these bacteria, which degrade dead plant and algae material, life itself would be impossible.) In that same teaspoon of water, however, are about ten times as many viruses.

Water is not unique in this way. A teaspoon of ordinary soil contains around 240 million bacteria and about 600 million viruses. (The entire continent of North America has fewer than 600 million people.)

In soil, bacteria are major players in the decomposition of organic matter and the cycling of chemical elements such as carbon and nitrogen, which are necessary for human life. Because plants can't create some of the nitrogen molecules they need to live, soil bacteria play an indispensable role in turning atmospheric nitrogen into the forms of nitrogen that plants need to survive. (The essential contributions of bacteria, fungi, and viruses to the ecological niches of the earth are discussed further in chapter 5.)

The estimated combined weight (or biomass) of all the plants and animals on Earth is 560 billion tons of organically bound carbon. According to a recent study published in the *Proceedings of the National Academy of Sciences*, researchers at the Weisman Institute in Israel and the California Institute of Technology determined that about 80 percent of the earth's biomass is composed of plant life.[2] And the second major component of Earth's biomass is bacteria (about 10^{30} of them), contributing about 15 percent to the global biomass. The biomass of fungi and archaea exceeds that of animals, and even more amazing is that the biomass of humans is surpassed by viruses.

Based on these calculations, germs are quite literally a weighty matter.

LIFESTYLES OF GERMS

To fully understand how beneficial germs can be, we need to look at ecosystems.

The term *ecosystem* was coined in 1930 by Roy Chapman in response to a request from the British botanist Arthur Tansley. Subsequently, Tansley became recognized as the founding father of ecology. He fully developed the concept of ecosystems as interacting communities of living organisms (plants, animals, and microbes) with nonliving components of their environments, such as air, water, and mineral soil.

A core concept of ecology is that, in nature, everything is connected. This concept was formulated by the brilliant German naturalist Alexander von Humboldt in the early 1800s, and some decades later was advanced by other visionaries, including environmental philosopher John Muir, who wrote, "When we try to pick out anything by itself, we find it hitched to everything else in the Universe."

The realization that human health is intertwined with the many living and nonliving components of our environment is a much more recent insight.

Planet Earth's first living species, our ancient microbial ancestors, were extremophiles (from the Latin *extremus*, meaning "extreme," and the Greek *philia*, meaning "love"). These microbes lived and reproduced in hostile environments, such as extreme heat, cold, acidity, and salinity.

Extremophiles continue to populate our planet today. They have been found living in the cold and dark, in a lake buried a half mile deep under the ice in Antarctica; in the deepest spot on Earth, at the bottom of the Mariana Trench in the Pacific Ocean; and inside rocks up to 1,900 feet *beneath* the sea floor, under 8,500 feet of ocean. And recently, a global community of more than one thousand scientists, named the Deep Carbon Observatory, released their findings that 70 percent of Earth's bacteria and archaea (fifteen to thirty billion tons of them) exists in the subsurface of our planet.[3]

In its infancy, life on Earth was hellish, with scorching temperatures way above the boiling point and an atmosphere composed of poisonous gases. The earliest microbes—like the archaeans living today—thrived without oxygen. (Such organisms are called anaerobes.) Billions of years ago, there was no oxygen in our planet's atmosphere. Fortunately, around 2.3 billion years ago, cyanobacteria began adding oxygen to the earth's atmosphere (called the Great Oxygenation Event), setting the stage for organisms that need this molecule (called aerobes)—including us humans. Nevertheless, there must have been an amazing variety of types of germs (bacteria, archaea, fungi, protists, and viruses) living together throughout the earth's watery environments way before animals emerged from the sea some 550 million years ago.

THE SOCIAL LIFE OF GERMS

Microbes don't usually exist as autonomous single-celled entities. In fact, many germs are highly social creatures that live in communities. Often these communities are polymicrobial—made up of multiple types of microbes. That cooperative interactions between individual bacteria within these communities occur may come as a surprise. But as noted by the American biologist E. O. Wilson, "Bacteria have thus been found to be social to a degree almost unimaginable to scientists a generation ago."[4]

In his 2018 book *The Strange Order of Things: Life, Feeling, and the Making of Cultures*,[5] neuroscientist Antonio Damasio traces the origins of the human mind and human cultures to the origin of life itself—to bacteria almost four billion years ago. Reflecting the book's title, he suggests that "strange" is too mild a word to describe this primordial connection. He observes,

> Bacteria are very intelligent creatures; that is the only way of saying it, even if their intelligence is not being guided by a mind with feelings and intentions and a conscious point of view. They can sense the conditions of their environment and react in ways advantageous to the continuation of their lives. These reactions include elaborate social behaviors. They can communicate among themselves. . . . There is no nervous system inside these single-celled organisms and no mind in the sense that we have. Yet they have varieties of perception, memory, communication, and social governance.

Bacterial cells communicate with one another by releasing chemical signals in a process called quorum sensing. And here's some more strangeness: Justin Silpe, a graduate student working in the Princeton laboratory where quorum sensing was discovered, recently found that bacteriophages—the viruses that infect bacteria—often eavesdrop on bacterial communication and use the information they pick up to harm those bacteria.

Like microbes, the vast majority of people are either harmless or beneficial to other members of *Homo sapiens*; only a small minority of people pose a threat to the well-being of others.[6] But look out when communication between members of our species is hacked by marauders, like the bacteriophages listening in to the quorum signaling of bacteria.

Like all other organisms, microbes are engaged in a constant struggle for existence. But it's not always a germ-eat-germ world. Some microbial communities have been shown to cooperate through metabolic cross-feeding, where one organism synthesizes a compound that another organism requires but can't produce.

Several kinds of such alliances are possible. As is the case for human relationships, some are good, some are bad, and others are indifferent. When two species live intimately together, their relationship is described as symbiotic. Symbiotic partnerships are a major source of evolutionary innovation. When the symbiotic relationship benefits both species, it is called mutualism. The relationship described initially by Lynn Margolis, of bacteria-derived mitochondria residing within eukaryotic cells, is referred to as endosymbiosis. When one member of the relationship benefits and the other neither gains nor loses, it is called commensalism. When one species benefits but the other pays a price in some way, the relationship is referred to as parasitism.

Some microbes are photosynthetic—they make their own food from sunlight, just like plants. Other germs absorb food from the material they live on or in. (The microbes that live in your gut absorb nutrients from the digested food you've eaten.) And some live off energy obtained from chemical reactions between water and rock.

Most bacteria reproduce by binary fission, a process in which an organism first duplicates its genetic material—its DNA—and then simply divides in two, with each new organism receiving one copy of the DNA. Binary fission is extremely efficient. If conditions are just right, one microbe can give rise to a billion progeny in only ten hours.

Germs have also evolved sophisticated strategies for defending themselves against competitors—in particular, by producing molecules that we humans have made into antibiotics. Penicillin, for example, which kills a wide variety of harmful bacteria, is released from a fungus, *Penicillium*. Recent studies indicate that bacteria also produce antibacterial peptides called toxins. And like antibiotics, toxins appear to hold promise as therapeutic agents.

But the same evolutionary strategies that gave us penicillin have also enabled microbes to evolve resistance to our antibiotics. (Much more on this in chapter 15.)

HOW GERMS GET AROUND

When European explorers first set foot in the Americas in the fifteenth and sixteenth centuries, they brought with them some highly contagious microbes—including the viruses that cause smallpox, measles, and influenza, and the bacterium that causes plague. It was these infectious agents—far more than guns and other weapons—that caused a massive population decline among Native Americans. (At least one well-known microbe—*Treponema pallidum*, the bacterium that causes syphilis—appears to have been transported in the opposite direction, from the New World back to Europe by returning explorers.)

Germs have developed multiple strategies for getting around. One major method is person-to-person spread. Since airplanes make over one hundred thousand flights each day, it's very easy for travelers to quickly carry microbes from one part of the world to another.

The folks who carry germs that cause disease, called carriers, may or may not get sick themselves. But they harbor pathogens on their hands or in their respiratory, gastrointestinal, or genital tracts. In the hospital, carriage of pathogens on the hands of healthcare providers or inanimate surfaces, such as stethoscopes and "scrubs," plays an important role in healthcare-associated infections.

Person-to-person dispersal of germs most commonly occurs through coughing or sneezing. This is how measles, tuberculosis, and the flu typically spread. Sexual contact is another common means of dispersal. That is how HIV, chlamydia, herpes, gonorrhea, and syphilis get from person to person.

Another common way for microbes to travel is by contaminating our food or water. One person with contaminated hands can easily spread germs to hundreds of others. Bacteria carried on our hands can be transferred to inanimate surfaces, and by this means they take rides on subways, ships, and airplanes.

In the biological world, genes generally travel vertically—that is, from what are called parent cells to daughter cells. But in the bacterial world, genes can also move between cells of unrelated species, through what is known as horizontal gene transfer (HGT). This feat is accomplished by the delivery of genetic material (DNA) packaged in bacteriophages, or in small pieces called plasmids. HGT allows bacteria to evolve at a blistering speed. As mentioned earlier, it also plays a key role in the emergence of bacterial resistance to antibiotics, as we'll see in chapter 15.

Lastly, certain germs hitchhike on animals or insects to get around. Some even hijack their hosts. More on this in later chapters.

3

THE HUMAN MICROBIOME

"You live in intimate association with bacteria, and you couldn't survive without them."—Bonnie Bassler

"We leave traces of ourselves wherever we go, on whatever we touch."—Lewis Thomas

From the early years of my career, I understood that most microbial species were either harmless or beneficial to human health. But back then I knew nothing about the groundbreaking research of Carl Woese and his colleagues at the University of Illinois. As noted in chapter 1, in 1977 these researchers described a whole new microbial domain of life: Archaea.[1] The technology they used to discover these microbes—and others like them that can't be grown (or, as scientists say, cultured) in the laboratory—is called metagenomics. This technology can be used to probe every nook and cranny of our planet—and even outer space. It has also revolutionized many aspects of human medicine.

The term *microbiome* is often attributed to the molecular biologist Joshua Lederberg, who in 2001 defined it as the ecological community of commensal, symbiotic, and pathogenic microorganisms that literally share our body space.[2] Simply put, the human microbiome refers to the microbes that inhabit the human body. Like your brain, your microbiome weighs about three pounds.

In 2008, the National Institutes of Health launched the highly successful five-year Human Microbiome Project (HMP). The goal of this massive project, which involved two hundred scientists from eighty research institutions, was to identify associations among the human microbiome, health, and disease. Two hundred and forty-two young, healthy adults were recruited for the project. On these volunteers, scientists investigated the microbiology of five body sites, each of which served as an ecosystem for germs: gut, skin, mouth, respiratory tract (lungs and nasal cavities), and vagina.

Their findings were astounding—most notably, that a case can be made that *Homo sapiens* evolved as a sophisticated transportation system for germs. Or, as the American journalist Michael Specter put it, "Germs are us."[3]

Consider these numbers. The human body contains an estimated 37.2 trillion cells, and the large intestine, where most of our microbiome lives, accommodates 39 trillion bacterial cells. The human genome, however, has only about 23,000 genes, but our microbiome is estimated to harbor between two and eight *million* unique genes. Thus, the genetic repertoire of the human microbiome is more than one hundred times greater than that of the human genome. In a very real sense, your body is 99 percent microbial. And, like our fingerprints and our genes, every individual's microbiome is unique.

In fact, many researchers regard the microbiome as a newly discovered essential organ of the human body.[4] But unlike your other organs, development of your microbiome didn't take place until birth, as you exited your mother's body via her birth canal (or through the skin in the case of a cesarean section).[5] You picked up a wide variety of germs on this short journey. These microbes took up residence very quickly and were soon joined by many other microbes that entered your body through the air you breathed, the milk and water you drank, and the many things you touched. By the age of three, your germ population resembled the one you now have as an adult.

Unlike your genome, your microbiome typically changes over time, based on your environment. These changes are a two-way street, however. Microbes move from your body to your living space, and vice versa, at incredible speed. In 2014 Jack Gilbert, a microbial ecologist at the University of Chicago and director of its new Microbiome Center (one of many popping up all over the United States), reported that when one young couple moved into a hotel room, within twenty-four hours the room was microbiologically identical to their home.[6]

YOU AND YOUR MICROBIOME

The HMP uncovered a wide range of correlations between germs and illnesses that were previously thought to not have microbial origins. Some of the most compelling correlations are with obesity, type 2 diabetes, inflammatory bowel diseases (Crohn's disease and ulcerative colitis), irritable bowel syndrome, cardiovascular disease, colon cancer, asthma, allergies, and autoimmune diseases such as multiple sclerosis and systemic lupus erythematosus.

Of course, correlation does not signify causation. But it does now appear that germs could be contributing factors for some of these illnesses. And many researchers are exploring the causal link between the composition of the microbiome and most, if not all, of these diseases. Publications describing links between the microbiome and almost any disease you can think of are appearing at breakneck speed. In the pre-microbiome stage of my career, I never witnessed anything like this in the fields of infectious diseases or microbiology. (For an excellent assessment of where the field stands, see the review by Jack Gilbert and his colleagues in 2018.)[7]

Given the extraordinary complexity of the microbiome, it is a deeply daunting task to decipher which of the many thousands of microbial species play a causative role in health or disease. Armed with advanced technologies, however, a number of unflinching scientists are working on this formidable task.

The work of Martin Blaser, the Henry Rutgers Chair of the Human Microbiome and Director of the Center for Advanced Biotechnology and Medicine at Rutgers University, on just one bacterium, *Helicobacter pylori*, revealed yet another level of complexity. Some strains of this bacterium are pathogens—they cause peptic ulcer disease and stomach cancer. But other strains appear to be mutualists—they protect us against asthma, hay fever, allergies, and gastroesophageal reflux disease (GERD). While the beneficial role of *H. pylori* has yet to be established, these findings suggest that aiming to eliminate a bacterial species from the gut microbiome could come at a price.

Given the rapidly increasing appreciation of the importance of the microbiome in health and disease, anything that alters its composition is cause for concern. What has so alarmed Martin Blaser—and many other scientists and healthcare professionals—is the misuse and overuse of antibiotics, and the effects such practices have on the human microbiome. Antibiotics are commonly prescribed to treat viral infections—yet they are completely ineffective against such infections. (They work quite well against bacterial infections, however.) As a result, the average American child receives three courses of antibiotics in the first two years of life—and an additional eight courses during the next eight years.

But even a short course of antibiotics can result in long-term shifts in their microbiome. One study discovered that children who received antibiotics before they turned six months old were more likely to be overweight as seven-year-olds. Another study showed that youngsters at age fifteen who had been prescribed antibiotics seven or more times in their childhood weighed about three pounds more than those who didn't take these medicines. And a recent analysis of thirty-two observational studies carried out by Dutch researchers indicated that antibiotic use during the first two years of life significantly increased the risk of developing hay fever and eczema in adulthood.

Antibiotics aren't the only way to disrupt the microbiome. A report in *Nature* in 2018 by Lisa Maier and colleagues indicated that almost 25 percent of one thousand marketed drugs tested inhibited the growth of at least one bacterial strain in the gut microbiome. (Antipsychotics were particularly strongly represented.)[8] Some studies suggest that cesarean deliveries encourage the growth of microbes from the mother's skin, rather than from the birth canal, in the baby's gut. And this change in microbiota may affect an infant's metabolism. A recent review of 163,796 births reported that children born by cesarean section were 48 percent more likely to be overweight or obese as adults than those delivered vaginally. Meanwhile, the percentage of births by cesarean section has climbed dramatically in recent years, rising to well over 30 percent in the United States and over 50 percent in Brazil, Egypt, and the Dominican Republic. (In 1970, the rate in the United States was 5.5 percent; in 1980, it was 16.5 percent.)

Understanding exactly where the microbiome comes from following birth is still in its infancy. Investigators from Baylor University reported in *Nature Medicine* in 2017 that there was no difference in the gut microbiome of babies born vaginally or by C-section. Likewise, research to date falls short of proving a causative role of a baby's microbiome at birth or infancy and altered health over time.

Unquestionably, the practices of hand washing, disinfection, sanitation, and keeping pathogens out of our food and water have saved many millions, if not billions, of lives. But some scientists now believe we have become *too* hygienic. Advocates of this idea propose that the lack of early childhood exposure to germs increases our later susceptibility to allergies by interfering with the normal development of immunity. Mounting evidence supports what is called the "hygiene hypothesis," that childhood exposure to germs helps the immune system to develop normally and not to overreact to substances that trigger allergies and asthma, which are immunological disorders.[9]

In a well-known study reported in 2016 in the *New England Journal of Medicine*, Michelle Stein and her colleagues compared the immune profiles of Amish children growing up on small single-family farms to Hutterite children, who are similar genetically but grow up on large, industrialized farms. The Amish living in an environment full of barnyard dust (and rich in germs) had significantly lower rates of asthma. It appeared that substances in the dust reprogrammed the immune cells of the Amish children, thus protecting them against asthma.[10]

Along this same line, a study reported in 2017 by Anita Kozyrskyj, a pediatric epidemiologist at the University of Alberta, showed that the microbiomes of babies from families with pets (mainly dogs) contained two types of microbes that were associated with lower risks of allergic disease and obesity.[11] For those with young children, here's yet another potential reason that dogs (and possibly other furry animals) are one of your best friends.

But the more it is studied, the more complex our understanding of the positive and negative roles of the microbiome gets. We are just beginning to learn about eubiosis, a healthy balance among all the microbes in our bodies, and dysbiosis, a microbial imbalance.

Thus far, most of the research into the human microbiome has focused on bacteria, and we've learned quite a lot about that limb of the Tree of Life. Relatively little, however, is known about the archaean microbes, which are also present in huge numbers in our bodies. We don't yet know what constitutes a healthy and harmonious relationship between bacteria and archaea—or even among different bacterial species.

And don't forget viruses. The number of viruses naturally present in a healthy human gut (called the gut virome) far exceeds the number of bacteria and archaea combined. Some of these viruses, called bacteriophages, work their way inside bacteria and in so doing knock them off. For example, a virus with the uncatchy name *crAssphage* appears to limit the growth of bacteria that have been linked to obesity and diabetes. [12]

And there's still more. Your gut microbiome also contains a fungiome or mycobiome, made up of over one hundred different species of fungal germs. Yet research on the impact of the fungiome on health and disease is still in its infancy.

OUR BODILY ECOSYSTEMS

Over an average person's lifetime, they will eat about thirty tons of food and drink about thirteen thousand gallons of water. How many germs will they swallow over that time? Certainly quadrillions—a quadrillion is one thousand trillion—and probably many more.

While it is true that, each year, one in six Americans gets sick from a food-borne infection, the overwhelming majority of germs we swallow go in one end and come out the other.

However, there are also many types of germs that spend their entire lives inside us. In fact, many of the over two thousand bacterial species that colonize the human gut stay inside us for decades. Many of these species are equally at home in the guts of other animals, such as the family dog.

Scientists are now intensely researching the effect of diet on the gut microbiome. Early studies suggest that food preservatives may be linked to weight gain and glucose intolerance (a sign of type 2 diabetes), via their impact on the microbiome. In her book *The Microbiome Solution: A Radical New Way to Heal Your Body from the Inside Out*,[13] gastroenterologist Robynne Chutkan introduces the "Live Dirty, Eat Clean Diet." She has teamed up with Elise Museles, a certified eating psychology and nutrition expert, who provides recipes that are aimed at building a healthy microbiome.

Age alters our microbiome, too. In general, older people display a reduced diversity in their microbiota than younger adults. But centenarians (people who live to be one hundred or older) have microbiomes that are more diverse than those of younger elderly people. A large cross-sectional study conducted by a group of Chinese researchers suggests that the microbiomes of healthy seniors are similar to those of healthy young people.[14] (More about the potential contribution of the microbiome to longevity in chapter 17.)

Research on twins indicates that genetic factors also influence the gut microbiome, which in turn influences other aspects of our health. In fact, one recent study suggests that the kind of bacteria you have in your gut may affect your body weight. Researchers at Washington University transplanted microbes from obese twins into a group of germ-free mice, and microbes from lean twins into a second group. The mice that received microbes from obese twins gained more weight, even though they didn't eat any more than the mice that were given germs from lean twins.[15]

In a study from China published in *Nature Medicine* in 2017, the bacterium *Bacteroides thetaiotaomicron* was found to be depleted in stool samples from obese subjects. And when this microbe was given to mice it prevented diet-induced obesity. Also, restoration of the abundance of *B. thetaiotaomicron* was seen in obese patients who had undergone bariatric surgery to treat their obesity.[16] But will a microbial transplant eventually enable people to stay or become thin? Many of the couple billion people who are overweight or obese—more than 44 percent of the global population—would love to be offered such a treatment.

One of the most exciting areas of recent research on the gut microbiome is occurring in the field of oncology. A growing number of studies have linked the composition of the gut microbiota to colorectal cancer. Also, connections between the gut microbiome and liver cancer, pancreatic cancer, and childhood leukemia have been reported.

It also appears that manipulating the gut microbiome may improve the effects of some of the most promising forms of immunotherapy. Researchers from the United States and France found that the composition of the gut microbiome can influence an individual's response to "immune checkpoint inhibitors." Immune checkpoint inhibitors release the breaks (i.e., checkpoints) on the immune system to activate it against cancer.[17] This could be a game changer because immunotherapy appears to have the potential to treat highly malignant cancers such as metastatic melanoma.

The microbiome of the skin is a second ecosystem that is under intense study. The skin is the largest organ in the human body. An average-sized adult's skin weighs about twenty pounds and has a surface area of around twenty square feet—but it is only a tenth of an inch thick. Your skin provides a watertight shield, exudes a variety of antimicrobial substances that kill or protect against pathogens, and plays a key role in vitamin D metabolism.

Because of the HMP, we now know that the skin is an extraordinarily rich ecosystem. Although inhabited by far fewer bacteria than the gut, the skin is nonetheless home to about one thousand bacterial species, as well as hundreds of species of fungi. (The heel harbors the greatest fungal diversity, with about eighty species, sixty of which can also be found in toenail clippings.)

Your skin is not as crowded with germs as your gut; it only has a trillion or so microbial inhabitants. As with your gut, however, almost all the microbes on your skin are either harmless or helpful. Also, as in the gut, the microbiome of each person's skin is unique. A recent study by researchers at the University of Waterloo of skin swabs from seventeen parts of the body of ten sexually active couples showed that each person significantly influenced the microbial communities on a lover's skin.

We are only now beginning to understand what the microbial diversity of the skin means for skin diseases such as acne, atopic eczema, psoriasis, rosacea, and skin cancer. (The microbiome of our gut may play a role in many of these illnesses as well.) Emma Barnard and her colleagues at the Geffen School of Medicine in Los Angeles suggest that it is the overall balance of the microbes in the skin microbiome, rather than the presence of individual bacterial species, that leads to skin health or the development of acne.[18] They also propose that insights from this kind of research may lead to more effective treatments of skin diseases with probiotics and bacteriophages. (More on these germ-based therapies in chapters 17 and 18.)

Studies of the microbiota of skin in the human armpit by Chris Callewaert at the University of California, San Diego, are aimed at determining the types of bacteria that give people offensive underarm odors.[19] Based on his findings, a new field of therapy, called armpit microbiome transfer, is in its early stage of development. (A far more advanced form of microbiome transplantation, fecal microbiota transplantation, is the topic of chapter 16.)

A third microbial ecosystem, your mouth, harbors a diverse microbiome, including a wide assortment of bacteria, archaea, viruses, fungi, and protists. So far, researchers have identified around one thousand bacterial species in the human mouth. Different areas—the teeth, gums, palate, back of the mouth—form niches for a variety of different long-term microbial residents. These microbes commonly live in biofilms, colonies firmly attached to surfaces of the mouth. Millions of minuscule creatures live together in these colonies, under a protective layer that keeps out potential invaders.

Almost everyone knows about two common bacterial infections of the mouth—cavities (technically known as dental caries) and periodontitis, an infection of the tissue surrounding the teeth. Few people realize, however, that microbes in the mouth have been implicated in cardiovascular disease, pancreatic cancer, colorectal cancer, rheumatoid arthritis, and preterm birth, as well as cancers of the head and neck. Researchers are now working to decipher just how and why these inhabitants of the mouth have such widespread influences throughout the body.

Amazing findings are also coming from studies of the forth microbiome: the lungs and sinuses. Our lungs are kept clean through millions of tiny hairlike structures called cilia, which line the respiratory tract. These cilia push particulates up and out of our respiratory system, either into our mouths or out into the world.

When I was in medical school, I was taught that these cilia kept the lungs largely sterile. But scientists have recently learned that our lungs aren't sterile at all. Although far less populated by microbes than our mouth or gut, a disease-free lung is inhabited by a persistent community of bacteria, as well as by archaea, viruses (including some helpful ones), and eukaryota (including some fungi). Healthy lungs also normally harbor colonies of *Penicillium*, the mold that produces the antibiotic penicillin.

Further adding to an appreciation of the complexity as well as the wonderment of the human microbiome are the recent findings of Robert Dickson and his collaborators showing that there is a connection between the microbiomes of the gut and the lungs that appears to contribute to lung health and disease. Based upon concepts regarding the so-called gut–lung axis, clinical trials are underway to see if manipulating gut bacteria can influence lung health.[20]

The microbiome of the vagina is the fifth ecosystem characterized in the HMP. For those of us born vaginally, the birth canal is where we were introduced in a major way to the microbial world. Many of the germs we picked up on this journey help to keep us healthy for the rest of our lives. Much of the recent scientific attention on the vaginal microbiome has focused on a bacterium called *Lactobacillus*. There are over eighty different species of *Lactobacillus*, including some that are found in yogurt. *Lactobacillus* is an especially helpful microbe because it produces lactic acid and hydrogen peroxide, both of which are toxic to potentially harmful microbial competitors.

Nearly a third of American woman have an infection known as bacterial vaginosis, or BV, which is associated with an increased risk of HIV, gonorrhea, chlamydia, pelvic inflammatory disease, and preterm birth, which is the leading cause of infant mortality.

Many factors affect the vaginal microbiome, including smoking, stress, diet, obesity, and the number of sexual partners. One of the most direct ways to alter the vaginal ecosystem is by douching. Although considered by many women a hygienic practice, because douching adversely affects the vaginal microbiome many authorities strongly discourage it.

THE GUT–BRAIN CONNECTION

From an evolutionary perspective, it makes sense that your gut microbiome communicates with your brain. After all, microbes inhabited Earth for several billion years before taking up residence in mammals, including *Homo sapiens*. And all organisms need to eat.

Germs in your gut are nourished by the food you consume, and animal studies suggest that gut bacteria may actually influence your food choices. While your brain itself is uniquely protected from germs, and appears to want nothing to do with a microbiome of its own, preliminary studies reported at the *Neuroscience* annual meeting in 2018 showed evidence by high-resolution microscopy of bacteria inhabiting cells of healthy brains.

Research in my neuroimmunology laboratory was focused for more than twenty years on defense systems of the brain. If the findings of bacteria in healthy brains are confirmed by others, it will not only blow my mind and those of most neuroscientists, but it will open a whole new chapter on brain diseases for which there are currently no known causes.

What we know now, however, is that there is a rich route of communication between your gut and your brain via your autonomic nervous system. For example, neurochemical signals released by nerves connected to your gut can affect your mood, making you happier or less happy, relaxed or anxious, sleepy or alert, hungry or full. More than 50 percent of dopamine and serotonin, the body's natural mood enhancers, are produced in the gut. All of this communication goes on automatically and unconsciously, 24/7.

How much do microbes shape human neural development, behavior, and brain diseases? Science is only now beginning to seek answers to these important questions. What role, for example, does our gut microbiome play in human cognition, sleep, mood, eating disorders, mood disorders, and poorly understood illnesses such as chronic fatigue syndrome (also called systemic exertion intolerance disease) and autism? (People with autism are much more likely than other folks to have gastrointestinal problems.) Furthermore, studies reported in 2016 by California Institute of Technology scientist Timothy Sampson and his colleagues using a mouse model of Parkinson's disease suggest that the gut microbiome may play a role in this neurodegenerative disease.[21] And their findings implicated activation of microglia, cells of the immune system that reside in the brain, in damage to neurons.

Emeran Mayer, a UCLA neuroscientist and gastroenterologist, suggests in his recent book *The Mind-Gut Connection: How the Hidden Conversation within Our Bodies Impacts Our Mood, Our Choices, and Our Overall Health*,[22] that "the connection between our gut and our mind is not something that solely psychologists should be interested in; it's not just in our heads."

Some of the most provocative evidence that gut bacteria can influence emotions has emerged from studies in mouse models of depression. In one study, the bacterium *Lactobacillus*, which is typically found in yogurt, was found to play a critical role in modulating metabolites associated with depression.[23] (You will read more about probiotics like yogurt in chapter 17.) And such animal studies underlie the thinking behind attempts to treat depression in humans with fecal microbiota transplants (the subject of chapter 16).[24]

Currently, most of the claims regarding how your gut microbiome can affect your emotional as well as physical well-being are extrapolated from studies in animals. With increasing recognition of the potential role of the human microbiome in health and in diseases for which improved treatments are desperately needed, profiling of the gut microbiome is becoming big business. But as underscored by Susan Lynch and Oluf Pedersen in their 2016 *New England Journal of Medicine* review, "The Human Intestinal Microbiome in Health and Disease," we need to be cautious until evidence is provided by properly controlled studies in humans.[25] Nonetheless, the potential of applying microbiome science to the discovery of new medicines has captured the attention of many research groups, as well as pharmaceutical companies.

In the arena of precision medicine or personalized medicine, which focuses on patients as unique individuals with unique genomes, the challenge now is to incorporate the much larger genome of your microbiome into therapeutic strategies. Pointing to the future, Rodney Dietert, a professor of immunopharmacology at Cornell University, forecasts that "precision medicine for the superorganism will treat you like an ecosystem. All of your body's thousands of species on the skin and in the gut, mouth, nose, airways, and reproductive tract need to be included within your health management."[26] How's that for an optimistic vision?

You've heard of blood banks. But you may be unaware of international efforts to bank fecal samples. Such "stool banks," with deposits from many ethnic groups around the world, could prove crucial in preserving the biodiversity of the gut microbiome, which has been markedly disturbed by modern life. It is hoped that these stool samples will someday pay off with discoveries of new treatments for many illnesses.

While we humans tend to be overly anthropocentric, focusing mainly on the human microbiome, many research groups are probing the microbiomes of other animals and plants, as well as many inanimate environments that we interact with on a daily basis, such as human homes, other buildings, subways, airplanes, and more. (The full scope of microbiome research is covered in Rob Dunn's excellent book, *The Wild Life of Our Bodies: Predators, Parasites, and Partners That Shape Who We Are Today*.)[27]

4

DEPARTMENTS OF BODILY DEFENSE

"The supreme art of war is to subdue the enemy without fighting."—
Sun Tzu

We can thank a family vacation in Messina, Sicily, for one of the most brilliant biological discoveries of all time. It was there in 1882, while his family was off at the circus, that Elie Metchnikoff, a Ukrainian zoologist, inserted thorns from a tangerine tree into transparent starfish larvae. The next day he observed with his microscope that cells surrounded and engulfed the splinters.

What Metchnikoff witnessed was the cells of the larvae's immune system accumulating at the site of the injury. You've likely seen this phenomenon many times with your own body, when a thorn or splinter pierced your skin, giving rise to inflammation—redness, swelling, warmth, and pain.

During that vacation, the idea popped into Metchnikoff's head that these cells—later named phagocytes (from the Greek *phago*, meaning "eating," and *cytes*, meaning "cells")—could play a critical role in defending against foreign invaders, especially bacteria.

Metchnikoff's observation in Messina came the same year that Robert Koch discovered the bacterium that causes tuberculosis. Metchnikoff's research led to an understanding of how the body defends itself against that disease, through what is now called cell-mediated immunity. (In 1908 Metchnikoff was awarded the Nobel Prize in Physiology or Medicine for his work on immunity.)

In 1888, Metchnikoff began working at the Pasteur Institute in Paris. By then, Louis Pasteur had already helped to develop the germ theory of disease. Along with Mechnikoff and other contemporaries, Pasteur also played a pivotal role in understanding the immune system.

One of Pasteur's best-known contributions to this field occurred in 1885, when he vaccinated a nine-year-old boy, who had been bitten multiple times by a rabid dog, with a weakened strain of the rabies virus. The vaccine prevented the boy from contracting rabies—an infection that, even today, is almost always fatal. (The extraordinary development of vaccination had been ushered in almost a century earlier by Edward Jenner after he inoculated a thirteen-year-old boy with vaccinia virus and demonstrated immunity to smallpox. More on this topic in chapter 6.)

THE BIG QUESTIONS

Over the past century and a half, the questions in immunology that have captured many minds (and several Nobel Prizes) have been these: How do cells of the immune system tell the difference between the body's own cells and foreign cells? And, in the case of germs, how do they discern between those that are dangerous and those that are helpful or benign? The answers are turning out to be complex indeed. Scientists have solved part of this mystery, but by no means all of it.

Another essential question has been this: Can our immune system harm us as well as protect us? We now know that the answer is definitely yes.

But first, what is the immune system? Like other bodily systems, the immune system is a network of cells, tissues, and organs that work collectively to defend our bodies against attack by foreign invaders, that is to say, germs. Intriguingly, cells of the immune system also play an important role in eliminating cells of the body that don't belong there (cancer cells). Recent studies by Felix Meissner and colleagues revealed that the various immune cell types form a social network,[1] much the same way that we are learning many microbial communities behave.

The principal cell types of the immune system (lymphocytes, macrophages, and neutrophils) have the absolutely uncanny ability to recognize foreign cells (germs and cancer cells) by the constituents on their surface.

It's something like you picking out wholesome versus spoiled fruit and vegetables before eating them.

Essential for survival, the immune system is a double-edged sword. Most of the time, when we become infected by a pathogen, it's not the infection itself that makes us sick or kills us. Cells of our immune system respond by releasing proteins called cytokines. These proteins travel to the brain, where they trigger symptoms of infection, such as fever, loss of appetite, fatigue, and aches and pains. Up to a point, those symptoms can be helpful, because they force us to slow down and take it easy—and, perhaps, crawl into bed and sleep. But if the immune system overreacts, that overresponse can kill us—as it did with tens of millions of people in the 1918 flu epidemic.

Our immune system can harm us in another way. If it loses its ability to distinguish its own cells and inadvertently attacks them, this becomes the cause of an autoimmune disease, such as multiple sclerosis, rheumatoid arthritis, or lupus.

Pathogens—the germs that cause disease—are often quite clever. They can rapidly evolve and adapt to avoid detection by the immune system. In response, animals' immune systems have coevolved multiple defense mechanisms to recognize and neutralize them.[2]

One such mechanism is called the adaptive immune system, in which certain cells (called B and T lymphocytes) have the remarkable ability to remember dangerous microbes they have encountered before. As soon as they recognize familiar and dangerous bacteria, viruses, fungi, or parasites, they quickly eliminate these enemies. Immunization (vaccination) works by stimulating the adaptive immune system.

One of the most exciting areas of immunology relates to the gut microbiome you read about in the preceding chapter. Recent evidence suggests that the microbes in your gut can affect (or even control) the development of your adaptive immune system. Our gut microbiome (a combination of bacteria, viruses, fungi, and other organisms) educates our immune system, teaching it how to tell friend from foe.

The adaptive immune system, which is centered on lymphocytes, emerged approximately five hundred million years ago, about the same time that the nervous system evolved in vertebrates. Adaptive immunity—the training of lymphocytes—takes time to develop. To fight infection from the moment of first contact, your body uses a more ancient mechanism, called innate immunity. Three types of cells in the innate immune system—neutrophils, macrophages, and natural killer (NK) cells—immediately sense and attack any potential pathogen. This counterattack is what creates inflammation.

Here are the three most important things to remember about the human immune system:

1.

    a. It evolved to protect us against microbes that breach our first line of defense (our skin, the lining of our gut, etc.).

    b. It is made up of four kinds of highly specialized cells: neutrophils, B lymphocytes, macrophages, and T lymphocytes. In later chapters, we'll look more closely at how these different types of cells protect us.

    c. If a defect (called an immunodeficiency) occurs in one of these types of cells, then the germs that pose the biggest threat will be those that are ordinarily contained or destroyed by that part of our bodily defense system. These germs are aptly named opportunists. Common types of immunodeficiency develop in those taking medications that impair the function of cells of the immune system. These medications include cancer-fighting agents (some of which wipe out immune cells residing in the bone marrow), drugs that prevent rejection of organs following transplantation, and medications that dampen the inflammation associated with autoimmune diseases. Also, extremes of age are accompanied by immunodeficiency—infants haven't had sufficient time to develop adaptive immunity, and like many other bodily systems, functioning of the immune system can wane in elderly individuals.

Only in recent years have we begun to deeply understand and appreciate how the human microbiome actually trains our immune system. However, we have also begun to recognize other, less positive aspects of our microbiome. For example, research at the Weizmann Institute in Israel, reported in 2017, suggests that some bacteria from our microbiome contain an enzyme that blocks the healing power of a common drug used to treat cancer.[3] I often reminded medical students in discussions of the immune system that the work of their tireless one-celled internal troops goes on day and night, without our knowing it. And I suggested that each night, before going to bed, they take a moment to thank their neutrophils, B and T lymphocytes, and macrophages.

Your body has yet another form of protection. Like the Great Wall of China, it is designed to keep out invaders. Its surfaces—your skin, your gastrointestinal tract (from the tip of your tongue to the end of your anus), your respiratory tract (from your nose and sinuses into the deepest parts of your lungs), and your genitourinary tract (tubes to and from your bladder and reproductive organs)—are all lined by barrier cells called epithelial cells. Just the lining of your gut holds forty trillion bacteria in place, separated from the rest of your body. The epithelial cells lining the colon (colonocytes), however, turn out to do a lot more than just provide a physical barrier to microbes.

For the most part, epithelial cells are very effective at keeping microbes from entering the bloodstream. But nature isn't perfect, and pathogens are very sly, so from time to time germs get across this barrier. That is why we have an immune system.

5

IT'S ALL CONNECTED

The Health of Humans, Animals, and Our Planet

"It cannot be said too often: all life is one. That is, and I suspect will forever prove to be, the most profound true statement there is."—Bill Bryson

Let us stop for a moment and recall the original concept of ecosystems provided more than eighty years ago by Sir Arthur George Tansley: interacting communities of living organisms (plants, animals, and microbes) with nonliving components of their environment, such as, air, water, and mineral soil.

In the twenty-first century, this definition has evolved into what is known as the One Health perspective.[1] One Health—at once a goal, a mandate, and a requirement for human survival and well-being—is typically defined as the collaborative effort of multiple disciplines—working locally, nationally, and globally—to attain optimal health for people, animals, plants, and our environment.

As recognized by Tansley, healthy ecosystems (environments) include a huge variety of microbes along with other living organisms as well as crucial inanimate constituents, like water and air. As mentioned in chapter 3, studies of human ecosystems (our gut, skin, mouth, respiratory tract, and vagina) have revealed an extraordinary array of helpful microbes that live and thrive on and inside us. If we don't recognize their value, support their efforts, and otherwise treat them with respect, we may endanger our species—and our world.

And it's not just the microbes that live inside us. Many external germs are hugely beneficial to human life as well. As just one example, consider the work of Pius Floris in Spain.[2] Floris's company has applied beneficial microbes—in this case, a species of fungi—to unproductive fields in the Castile and León region. These fungi are beginning to return the soil to a productive state. As Floris explained, "Farmers have ignored these symbionts for decades. We are bringing them back into the game."

In similar fashion, humans are now employing technologies that harness the beneficial germs in plant roots, called rhizobiomes. The environmental toxicologist Emily Monosson recognizes rhizobiomes as the vegetal equivalent of our gut microbiome. In her book *Natural Defense: Enlisting Bugs and Germs to Protect Our Food and Health*,[3] she warns that "wholesale destruction of bacteria, whether in the human body or the agricultural field, can be profoundly disruptive."

Along with the field of evolution, environmental science has accelerated remarkably in the past several decades. While the human microbiome has captured most of the scientific and popular attention, similar studies have been underway aimed at characterizing the microbiomes of many other animals, plants, soil, and water. As just one of many examples, a recent report by Nancy Moran and her colleagues at Yale University suggests that the collapse of honeybee hives in recent years is related in part to overuse of antibiotics in agriculture that selects antibiotic-resistant, deleterious bacteria in the gut microbiome of the honeybees.[4] (This finding is reminiscent of those of Martin Blaser and others that raised concern about overuse of antibiotics in humans, mentioned in chapter 3.)

Also, a multitude of studies of soil and seawater in the past several decades have shed increased light on the incredibly important role of beneficial bacteria, viruses, and fungi in vital processes such as nitrogen fixation, nutrient recycling, biodegradation, production of oxygen, and removal of carbon dioxide from the atmosphere.

I suspect that, within the next decade, microbes will be more noticeably factored into our definition of One Health given their crucial role in human, animal, plant, and planetary health.

One of the most extraordinary aspects of the One Health movement is its multidisciplinary nature. The University of Minnesota, where I've been on the faculty for four decades, offers a good example. Its One Health program (also called One Medicine, One Science) involves the College of Veterinary Medicine; School of Public Health; Medical

School; College of Food, Agricultural and Natural Resource Sciences; School of Nursing; Center for Animal Health and Food Safety; Center for Global Health and Social Responsibility; College of Science and Engineering; and Institute on the Environment. Everyone involved recognizes that if any one sphere of health—human, animal, plant, microbial, or environmental—goes down, we all go down.

The multidisciplinary approach of the global One Health initiative is underscored by the diversity of its financial supporters, including the United States Centers for Disease Control and Prevention, the Wildlife Conservation Society, the Food and Agriculture Organization of the United Nations, the World Bank, and UNICEF.

STRESS, EVOLUTION, AND ONE HEALTH

> "If you ask what is the single most important key to longevity, I would have to say it is avoiding worry, stress, and tension. And if you didn't ask me, I'd still have to say it."—George Burns

The endocrinologist Hans Selye first used the term *stress* in a biological context in 1936. The term had been used for centuries in physics, in relation to a material's ability to resume its original shape after being compressed or stretched by an external force. Selye defined biological stress as the nonspecific response of the body to any demand or change.

From the perspective of evolution, stressful environmental conditions have been major drivers of adaptation and the selection of fit species from the very origin of life. (Once again, consider the archaea and other groups of germs that continue to survive in hellish environments. And most members of the human microbiome as well as other microbiomes of our planet are composed of intimate friends that adapted to what we might consider rather gruesome environmental conditions, like the human gut, where about forty trillion germs thrive.)

In the early 1980s, the field of psychoneuroimmunology—an interdisciplinary field of research focused on interactions among the brain, the immune system, and the endocrine system—began to blossom.[5] Early research in this field provided clear-cut evidence that stress negatively impacts the immune system of humans and other animals. In laboratory studies, animals were subjected to a variety of stressors—cold, physical confinement, loud noises, and mild electric shocks. Animals exposed to these stressors developed more serious infections when they were challenged by microbes. Similar studies done with human beings under great stress—students taking final exams and people caring for patients with Alzheimer's disease—revealed a similar suppression of the immune system.

Remarkably, the bodies of all vertebrates—from lizards to goldfish to leopards to humans—respond to stress by secreting the same, or similar, hormones. Peptides similar to these hormones are also found in snakes, and even in invertebrates, such as insects, mollusks, and marine worms.

From a One Health perspective, a stressor that threatens the existence of *any* living organism—whether an animal, a plant, or a helpful form of microbial life—represents a threat to us all. And by "threat," I mean something that can permanently wipe species off the face of the earth. As you'll discover in chapter 20, of the thirty-six billion species that ever existed on Earth, 99 percent are now extinct. For one reason or another, they were literally stressed out. You will also read in that chapter about a colossal stressor that annihilated many of these species and that again threatens all living species, namely climate change.

Part Two

Mortal Enemies

6

WHAT HAS PLAGUED US?

"Humanity has but three great enemies: fever, famine, and war; of these by far the most terrible is fever."—Sir William Osler, founding professor of Johns Hopkins Hospital

IT'S AN INFECTIOUS WORLD

Just what, exactly, is an infection, and is it the same thing as an infectious disease?

You may be surprised to learn that infectious disease experts have not yet settled these questions. They have multiple views—and many debates—on the subject.

That said, here are some commonly accepted definitions, which are the ones I believe are the most helpful—and the ones I'll use in this book:

An *infection* is any established relationship between a microbe and a host. Thus, all the time, you are infected—some would say colonized—literally head to toe, and from the tip of your tongue to the other end of your gastrointestinal tract. In 99+ percent of these cases, these infections, caused by friendly or benign microbes (our intimate friends), don't give you any trouble.

But when an infection makes you sick, it is caused by a pathogen—a harmful microbe. In that case, you have an *infectious disease*. In contrast to those that simply colonize our bodily surfaces, bacterial pathogens produce what are called virulence factors, such as toxins that injure or kill cells of the host or that allow invasion of tissues that are otherwise off limits.

For the past several decades, it has become almost impossible to pick up a newspaper without reading about a new infectious disease epidemic—Legionnaires' disease, Lyme disease, HIV/AIDS, SARS, "flesh-eating bacteria," hepatitis C, West Nile virus encephalitis, bird flu, Ebola, and Zika virus infection, to name just a few. These new diseases are known as emerging infections. In fact, as was mentioned in the introduction, in 1992 the United States Institute of Medicine (IOM) published a landmark book called *Emerging Infections: Microbial Threats to Health in the United States*,[1] intended as a wake-up call for the U.S. Congress to take action.

The IOM's definition of an emerging infection is still used today: an infectious disease that has newly appeared in a population or that has been known for some time but is rapidly increasing in incidence or geographic range.

By the early 1990s, the onslaught of new or reemerging infectious diseases was staggering. To help physicians, hospital nurses involved in infection control, and public health practitioners stay abreast of new developments, my colleague, Michael Osterholm, and I initiated an annual course in 1992, "Emerging Infections in Clinical Practice and Public Health," cosponsored by the Minnesota Department of Health and the University of Minnesota. Now in its twenty-fifth year, the course continues to attract over three hundred attendees annually.

So what happened in the last quarter of the twentieth century to give rise to so many emerging (and often deadly) infectious diseases? Who or what tipped the balance in favor of our mortal enemies?

Perhaps unsurprisingly, the answer is *Homo sapiens*. The main underlying factor in most emerging infections is human behavior, or misbehavior.

Perhaps the biggest contributor is the extraordinary acceleration in transportation of people and food by planes. There is almost no better way to spread germs far, wide, and rapidly. Other human behaviors that play a role in the emergence of infectious diseases include unsafe sex, urbanization, deforestation, pollution of water and air, and political turmoil.

More than 60 percent of the estimated 140 emerging infections in humans are transmitted to us from animals. This means that physicians and veterinarians need to work together. Fortunately, more and more, they are doing so.

In the pages to come, we'll zoom in and look closely at some of the most instructive emerging infections that pose the biggest threats to humans. We'll also look at the most promising—and often surprising— ways to avoid these infections, keep them at bay, and cure them.

But before we look at modern emerging infectious diseases, let's consider the history of the most remarkable epidemics that occurred before the end of the twentieth century.

Until World War II, infectious diseases killed more combatants than weapons. And until the late twentieth century, infectious disease epidemics killed more people globally than cardiovascular disease and cancer combined. In terms of sheer carnage, epidemics are humankind's worst and biggest enemy.

But what exactly is an epidemic—also sometimes called a plague?

The *dem* in *epidemic* comes from the ancient Greek word *demos*, which meant "people" or "district." When an infectious disease sickens or kills a large number of people across a wide area, that disease is an epidemic.

If the infectious disease is restricted to a particular group in a defined geographic area, it is instead said to be endemic. And when an epidemic crosses international borders, the situation is called a pandemic.

In modern times, the terms *epidemic* and *pandemic* have been extended to cover many noninfectious diseases and harmful behaviors, such as obesity, heart attacks, type 2 diabetes, high blood pressure, cancer, drug abuse, and violence.

The word *plague* was derived from the Latin *plaga*, "pestilence." The Middle English word *plage* was coined during the fourteenth century, when bubonic plague was decimating Britain.

Initially, the term *plague* was used to describe that one specific epidemic, which was later found to be caused by the bacterium *Yersinia pestis*. Eventually, however, *plague* was used for any widespread (and often fatal) epidemic, as well as for other highly destructive forces. (Think of a plague of locusts, or super-annoying people who should be avoided like the plague.)

Epidemics have dramatically shaped the course of human history for thousands of years. The microbes that cause plague, smallpox, influenza, measles, and salmonella gastroenteritis—carried by European explorers to the New World—were the primary cause of Europeans' swift military victories over Native America. But Europeans also died by the millions in many dozens of plagues over the centuries.

In the Plague of Justinian, during the years 541–542, bubonic plague killed about 40 percent of all people in Europe. Since then, written records attest to at least 186 other epidemics. Of these, bubonic plague accounts for twenty-six and smallpox for twenty-one. Also high on the list are cholera (34), yellow fever (15), and influenza (13). (Of course, since human beings didn't understand the causes of these infectious diseases until the late nineteenth century, the causes of epidemics before that time are surmised from their descriptions in written records.)

In the past half century, epidemics caused by newly recognized viruses have captured the world's attention. These have included several new strains of influenza virus, West Nile virus, dengue virus, chikungunya virus, Ebola virus, Zika virus, and most notably HIV. I'll discuss all of these diseases in later chapters.

PLAGUES THAT ALMOST CRUSHED HUMANITY

"One death is a tragedy; one million is a statistic."—Joseph Stalin

The Speckled Monster: Smallpox

Smallpox likely emerged about 10000 BCE, and smallpox epidemics were recorded throughout the ancient world. Evidence of smallpox was found in three-thousand-year-old Egyptian mummies, including the mummy of Pharaoh Ramses V.

The first unequivocal description of smallpox in Western Europe occurred in the year 581, when Bishop Gregory of Tours provided an accurate account of the characteristic symptoms and rash. Europe later became a hub from which smallpox spread to other parts of the world via explorers.

Thankfully, we appear to have eradicated smallpox from the planet—something we have not yet done with any other illness that's infectious to humans. The last death due to smallpox occurred in Somalia in 1977.

It's hard to imagine just how devastating this disease was. People with smallpox suffered with a high fever, head and body aches, and sometimes vomiting before a characteristic, disfiguring rash erupted. The *pox* part of smallpox is derived from the Latin word for "spotted," and refers to the raised bumps that appeared on people's faces and bodies. The disease—often called "the speckled monster"—was fatal in 20 to 60 percent of cases.

Smallpox was caused by the variola major virus, which was transmitted from one person to another via direct, and fairly prolonged, face-to-face contact. Even though it wasn't until the end of the nineteenth century that germs were understood to be a cause of disease—and it wasn't until 1906 that the variola major virus was identified—by the Middle Ages people had a sense that smallpox was contagious. When a case of smallpox appeared in town, many people fled. But some compassionate souls (among them family, clergy, and doctors) risked getting sick by staying put and caring for the ill and suffering.

During the eighteenth century, four hundred thousand Europeans were killed by smallpox each year, including five reigning monarchs. It is estimated that smallpox killed more people than all wars combined. And during the twentieth century, smallpox killed an estimated three hundred to five hundred million people worldwide—more people died from smallpox than from influenza, tuberculosis, HIV/AIDS, and malaria combined.

These statistics suggest that the eradication of smallpox was perhaps the single most important accomplishment in the entire history of medicine.

Of the many individuals and institutions that contributed to this achievement, most notable were an eighteenth-century English country doctor, Edward Jenner, who discovered an effective vaccine, and the World Health Organization's Smallpox Eradication Program, carried out from 1966 through 1980.

Long before Jenner, people understood from simple observation that anyone who recovered from a bout of smallpox was resistant when reexposed to it. In China as early as the fifteenth century (as well as sporadically in Europe) people used a practice called variolation—the insertion of smallpox material, usually scabs, under people's skin—to immunize them against the deadly disease. While variolation often seemed to work, on occasion it resulted in people actually contracting smallpox.

In the early 1700s, the Puritan minister Cotton Mather was a strong proponent of variolation. At the time, it was a subject of intense controversy, and Mather's support of the practice made him many enemies. At one point a profane note, attached to a stink bomb, was thrown through a window into his house by an antivariolation zealot.

Things got more scientific toward the end of the 18th century. In one of the most classic experiments in the history of medicine, on May 14, 1796, Jenner used material from the sores of a milkmaid with cowpox to inoculate James Phipps, an eight-year-old boy. Jenner hypothesized that whatever caused cowpox, it was similar but far less virulent, and that it would likely prevent smallpox. When, two months later, James was exposed to smallpox sores, he did not fall ill. The vaccination with cowpox had worked.

Edward Jenner was also the first person to use the term *vaccine*, derived from the Latin *vacca*, meaning "cow." Thus Jenner became recognized as the father of vaccination. (However, Dr. Arthur Boylton makes a convincing case in his 2018 article "The Myth of the Milkmaid," in the *New England Journal of Medicine*, that the *idea* that cowpox could prevent smallpox infection actually originated in the mind of a different country doctor, John Fewster, in 1768.[2] To further tarnish the legend, an international group of researchers reported in 2017, also in the *New England Journal of Medicine*, that Jenner's vaccine may actually have been derived from a horse—that is, from horsepox virus, not cowpox virus.)[3]

But there is a cautionary note regarding smallpox: although it is considered eradicated, it is not extinct. By definition, *eradication* means a permanent reduction to zero cases worldwide, with medical intervention no longer required anywhere. *Extinction* means that the specific infectious agent no longer exists, either in nature or in the laboratory.

Samples of variola major virus are currently stored in super-secure laboratories in Atlanta and Russia. In 2001, following the bioterrorism threat of anthrax (see chapter 2) and rumors that the virus was one of the weapons of mass destruction harbored by Iraq, the terror of smallpox loomed again. As a result, in 2002, in preparation for a possible bioterrorism attack, I and many other healthcare workers received a booster smallpox vaccination.

Today, the National Institute of Allergy and Infectious Diseases still lists variola major as a Category A pathogen—an organism that poses the highest level of risk to national security and public health.

To date, attempts have been made to eradicate six other infectious diseases in humans. Although none has yet completely succeeded, in many cases the results have been excellent. The most promising outcome involves poliomyelitis, commonly known as polio. Because of a coordinated vaccination campaign initiated in the late twentieth century by the World Health Organization, the United Nations Children's Fund, and Rotary International, annual cases of polio plummeted by over 99.9 percent. In 2017, only twenty-two cases were reported globally, but in 2018 progress on eradicating polio stalled: by the end of November 2018, there were twenty-seven cases. And, tragically, in April 2019, Pakistani health officials suspended a nationwide antipolio campaign after a health worker and two policemen escorting vaccination teams were killed by militants. Nonetheless, many experts remain hopeful that, in years to come, polio will be eradicated from our planet.

The Black Death: Bubonic Plague

The devastation caused by smallpox is rivaled only by bubonic plague, which is fatal in 50 to 60 percent of all cases. The name of the disease derives from the ancient Greek word *boubon*, which refers to swelling in the groin by enlarged lymph glands, called buboes. This is one of the most evident symptoms of the illness; other common symptoms include fever, chills, diarrhea, and bleeding—typically from the mouth, nose, or rectum, or under the skin. Blackening and the death of tissue (known as necrosis) in the arms or legs occurs when the infection spreads to the bloodstream.

Unlike smallpox, plague is a bacterial infection, not a viral one. It is typically spread by animals.

The germ that causes bubonic plague, *Yersinia pestis*, was discovered in 1894 by Alexandre Yersin, a student of Louis Pasteur and Robert Koch. Yersin also found the bacterium in rats. We now know that many different rodents, from mice to prairie dogs, carry the disease, but rats are the most common carriers.

Yet bubonic plague is not spread *directly* from rats to humans. Fleas serve as intermediaries. First the rat gets infected; then a flea living on the rat bites the rat and gets infected as well; then the flea jumps from the rat to a human and bites them, thus transmitting the infection.

A recent study of bones from humans who died in the Bronze Age, some five thousand years ago, detected the DNA of *Y. pestis*. For centuries, travel along trade routes to and from China—and from port to port on ships carrying both humans and flea-laden rats—was the genesis of many plague epidemics.

There are actually three different kinds of plague. The most common form, which I described above, is bubonic plague. A second form, pneumonic plague, is a highly lethal infection of the lungs that can be spread by coughing. The third form, septicemic plague, occurs when the plague bacterium invades the bloodstream, and it is an almost certain death sentence. People dying of septicemic plague turn a very dark color—hence the name Black Death.

There have been twenty-eight epidemics of bubonic plague in recorded history, including the Great Plague of Athens (430–427 BCE); the Plague of Justinian (541–542), which killed twenty-five to fifty million people in the Eastern Roman Empire; the Black Death (1346–1353), which wiped out 30 to 60 percent of all humans in Europe; and the Great Plague of London (1665), which killed one hundred thousand Londoners—20 percent of the city's population—in seven months.

The Great Plague of London deserves our closer attention. Most healthcare practitioners responded to it by fleeing the city; only a small handful remained in London. Diarist Samuel Pepys, who also stayed in town, provided a graphic and terrifying description of the calamity. So did novelist Daniel Defoe in his *A Journal of the Plague Year*. Here is an excerpt:

> London might well be said to be all in tears . . . but the voice of mourners was truly heard in the streets. The shrieks of women and children at the windows and doors of their houses, where their dearest relations were perhaps dying, or just dead, were so frequent to be heard as we passed the streets, that it was enough to pierce the stoutest heart in the world to hear them. . . . Death was always before their eyes, that they did not so much concern themselves for the loss of their friends, expecting that themselves should be summoned the next hour.

Meanwhile, religious leaders decried the plague as God's punishment for people's sins—just as some of our televangelists do today.

But something else happened as well. For the first time, governments in Europe got seriously involved in medical matters. Public health boards were created. These organizations built pest houses for the sick and set and enforced strict quarantine measures.

At the time, of course, no one knew anything about germs. The contagion was thought to have been spread by poisonous fumes coming from various sources of "corruption." In practice, this meant that the poor and their surroundings were blamed as sources of the disease.

Cases of plague and deaths still occur today. In 2017, an outbreak of plague rapidly spread through Madagascar. More than two thousand citizens were sickened, and 165 died, before the outbreak was declared contained by the WHO in December.

However, because of the introduction of antibiotics, the death rate in the United States from plague fell from over 66 percent early in the twentieth century to 11 percent a century later. Also, nowadays plague outbreaks are promoted more by jet travel than by poverty and war.

In 1994, an outbreak of bubonic plague in India drew worldwide attention. During this outbreak, plane crews originating in India were required to notify health officials if a passenger on the plane was sick. If someone was, the plane was met by a health official upon landing, and the sick person was immediately quarantined.

When bubonic plague first came to the western United States a century ago, *Y. pestis* became established in wild rodent populations, including prairie dogs. As a result, about eight cases of bubonic plague are reported annually in the United States, all in the West.

In 2015, a small outbreak of human pneumonic plague occurred in Colorado. Four cases were reported, all of which appeared to have come from contact with an infected dog. In general, however, pneumonic plague is very rare in the United States; between 1900 and 2012, a total of only seventy-four cases were reported. In 2017, three cases of human plague were reported in Santa Fe County, New Mexico; in these cases, contact with prairie dogs, not dogs, was the source of infection.[4]

Because a plague vaccine isn't available, visitors to national parks in the western United States should protect themselves from bugs by spraying with repellant containing DEET, and should avoid feeding squirrels, chipmunks, and other rodents.

OTHER PLAGUES THAT CONTINUE TO KILL

The White Plague: Tuberculosis

The bacterium that causes tuberculosis, *Mycobacterium tuberculosis*, also known as the tubercle bacillus, is quite remarkable. It is transmitted exclusively person to person through coughing or sneezing. When a person gets infected, the germ can spread from the lungs to virtually every other organ.

But symptoms of tuberculosis, commonly known as TB, are actually relatively rare. About two weeks after someone becomes infected, their immune system kicks in. As a result, only 5 percent of infected people become sick. In the remaining 95 percent, the tubercle bacillus becomes dormant. This is called a latent infection.

You can be latently infected with tuberculosis and not know it. In fact, one out of every three people on Earth is latently infected with *M. tuberculosis*.

The great majority of these people will never develop any TB symptoms. However, if someone's immune system becomes impaired—through HIV infection, or treatment with certain kinds of medications, or old age—the tubercle bacillus can awaken and reactivate. The tubercle bacillus is essentially a zombie germ. The disease can then cause a variety of complications, depending on the organ it has quietly been residing in. For example, headache and a stiff neck occur in TB of the brain, abdominal pain is the main symptom when the tubercle bacillus reactivates in the abdominal cavity, and pain in the low back is common in patients with TB of the spine.

Like smallpox and plague, the impact of TB on human history has been enormous. During the nineteenth and early twentieth centuries, TB was called the Great White Plague—"White" because patients were usually pale from the anemia that TB creates, and "Plague" because, like the Black Plague that preceded it, TB was the cause of more deaths in industrialized countries than any other disease.

Social determinants such as poverty, discrimination, and overcrowding have typically played a large role in who contracts TB. Beginning in the twentieth century, as social conditions improved for many people, the rate of TB steadily declined. But when anti-TB antibiotics became available in the 1940s and 1950s, this was a total game changer. Suddenly patients didn't need to be quarantined in sanatoriums, sometimes for years. Instead, they could almost always be treated as outpatients. In the late 1960s, hundreds of sanatoriums around the United States closed, more or less overnight.

Recent studies of *M. tuberculosis* genetic material recovered from human skeletons in Africa and Peru suggest that humans acquired the disease in Africa about five thousand years ago. Our ancestors passed the disease on to goats, cows, and other domesticated animals. (This type of transfer is called anthroponosis. Infectious agents can also travel in the other direction, from animals to humans; this is known as zoonosis.) Later, it appears that infected sea lions and seals from Africa brought TB to the shores of South America. However, most of the evidence suggests that European explorers were responsible for spreading this illness to the New World.

Today, more than ten million people around the world come down with TB symptoms each year, and 1,800,000 people, mostly in developing countries, die of the disease. In fact, TB is currently the world's top infectious killer. Most of these deaths could have been prevented with antibiotics—but the sick and infected people simply live in countries that can't provide the infrastructure to treat and follow patients with TB.

One of the most alarming recent developments is the emergence of strains of the tubercle bacillus that are resistant to virtually all antibiotics. (More on this microbe in chapter 15.) The World Health Organization has established a fund to support research for containing this mortal enemy.

Today, the pharmaceutical industry, governments, and philanthropic organizations are all working on new vaccines and drugs. But the disease has not yet been eradicated. In part this is because, sadly, there isn't a lot of money to be made from anti-TB drugs, because most of the people who become ill with tuberculosis live in countries with a great deal of poverty. Nonetheless, the medical journal the *Lancet* recently released a report calling for increased investments in diagnosing, treating, and preventing TB, which could help end TB by 2045.[5]

Bad Air: Malaria

So far, my discussion of dangerous pathogens has dealt with two especially virulent prokaryotes, *Y. pestis* and *M. tuberculosis*. Both are members of the Bacteria domain of the Tree of Life.

The microbe that causes malaria, *Plasmodium*, is quite different. It's a single-celled member of the Eukarya domain of the Tree of Life. But don't be fooled by the fact that *Plasmodium* is more closely related to humans than it is to bacteria and viruses. Today malaria remains one of the biggest health threats to human beings. The main symptoms of malaria are fever, chills, headache, vomiting, diarrhea, and serious discomfort. Sometimes malaria is deadly.

The word *malaria* comes from the Italian words *mal*, meaning "bad," and *aria*, meaning "air." The ancient Romans blamed the air in swamps for the disease. They weren't too far off, as malaria is transmitted by the bite of infected *Anopheles* mosquitoes, which breed in swamps and other stagnant water.

This germ has caused enormous havoc throughout human history. Malaria appears in writings from ancient China, Egypt, and Greece, and it played a significant role in the fall of the Roman Empire.

It has also been a defining factor in many wars. During the Civil War, for example, more than one million Union troops contracted malaria, and roughly thirty thousand died. In the Pacific Theater of World War II, malaria was the single most common health hazard for U.S. troops: about five hundred thousand became infected.

In tropical countries where the disease was (and in many cases still is) rampant, malaria has greatly impeded human development. In Panama, for example, malaria, along with yellow fever virus, defeated France's attempts to build a canal in the 1860s. America's later success in canal building was made possible not because we were better engineers but because of public health measures developed and enforced by Walter Reed and William Gorgas. (Walter Reed was a U.S. Army pathologist and bacteriologist who helped prove that yellow fever is transmitted by the bite of a mosquito. The Walter Reed Hospital in Washington, DC, was named in his honor. William Gorgas was a U.S. Army surgeon who introduced mosquito control measures to prevent yellow fever and malaria.)

Malaria continues to be a serious problem today. In 2017, there were 219 million cases of malaria worldwide. To the relief of public health leaders, the death toll for that year had fallen to "only" 435,000, most of whom were children in Africa.

Two Nobel Prizes for Physiology or Medicine have been awarded for breakthroughs in understanding the genesis of the disease. In 1902, the

Scottish doctor Sir Ronald Ross was honored for his discovery of the complete life cycle of the parasite in mosquitoes. He also linked the disease to the bite of the female *Anopheles* mosquito. (Only female mosquitoes feed on blood; male mosquitoes feed on plant nectar and don't transmit the disease.) And in 1907, a French doctor, Charles Louis Alphonse Laveran, was recognized for his discovery that the parasite resides inside the red blood cells of infected people.

We now know that there are at least three thousand different species of mosquitoes in the world. Only 430 are species of *Anopheles* mosquitos, and of these only thirty to forty species—and only the females—transmit malaria.

We also know that there are about two hundred species of *Plasmodium*—and only five of these are responsible for virtually all the cases of malaria in humans. The most lethal species, *P. falciparum*, causes what is called malignant tertian malaria. Other *Plasmodium* species infect other animals—birds, rodents, reptiles, and nonhuman primates such as apes and chimpanzees.

Fortunately, several effective treatments are available. Quinine was the first effective antimalarial agent. Obtained by South American natives from the bark of the cinchona tree, quinine was among the most important discoveries of the Spanish conquistadors. By 1633, Jesuit priests had documented the ability of "Peruvian bark" to cure malaria, and quinine was in use in Rome, where malaria was endemic. Today quinine is still sometimes used to treat the disease—but it can be somewhat toxic, so artemisinin, which is derived from an herb used in traditional Chinese medicine, is the more common choice. In 2015, Dr. Tu Youyou, a member of the China Academy of Traditional Chinese Medicine, was awarded the Nobel Prize in Physiology or Medicine for her discovery of artemisinin.

Recent prevention measures, such as spraying with insecticides, draining stagnant water, and using insecticide-treated bed nets, have had a big positive impact. Since 2000, the rate of malaria infection has plunged by 60 percent worldwide. Former World Health Organization director-general Margaret Chan hailed this as one of the great public health success stories of our millennium.

In fact, malaria has now been eliminated in 111 countries, and thirty-four countries are advancing toward elimination. (As mentioned earlier, elimination is defined as "absence of disease in a defined geographic area." Eradication means worldwide elimination.)

Even though some of the world's brightest scientific minds have been working hard to develop an effective malaria vaccine—with financial support from government, philanthropy, and industry—they haven't yet succeeded, but significant progress is being made. In fact, a malaria vaccine that can provide up to 100 percent protection against the disease will be tested in a clinical trial in early 2020 on Bioko, an island off the coast of Equatorial Guinea.

The Blue Death: Cholera

Cholera is an infection of the small intestine caused by the bacterium *Vibrio cholerae*. The hallmark of the disease is copious, watery diarrhea. And by "copious," I mean three to five gallons a day.

Understandably, an untreated person with cholera will quickly become severely dehydrated. The dehydration can result in sunken eyes and wrinkling of the skin of the hands and feet, which can take on a bluish color—thus the nickname "the Blue Death."

In most cases, cholera can be successfully treated with proper oral rehydration therapy—large volumes of water supplemented with electrolytes. This is both inexpensive and easy to administer (people simply drink it). Literally millions of lives have been saved in recent years by this therapy. Nevertheless, each year three to five million people contract cholera worldwide, and between 55,000 and 130,000 die.

Cholera is all about water—either fresh water or brackish salt water—that has been contaminated by human feces. While we are the only complex animal that *V. cholerae* infects, cholera bacilli can live in water on a form of simple, microscopic animal life called zooplankton. Blooms of zooplankton typically go hand in hand with cholera outbreaks, especially in coastal areas of Southeast Asia.

Water containing *V. cholerae* can be siphoned by oysters and clams that eat zooplankton. This is why uncooked shellfish are a potential source of infection. (Because shellfish are more active in the warmest months—May through August—there is some validity to the folk wisdom that it is safer in the northern hemisphere to eat raw oysters and clams during months that contain an R.)

Viruses are another piece of the cholera puzzle. A common bacterio-phage—a virus that infects bacteria—incorporates itself into cholera bacteria. Once there, it hijacks the bacteria and directs it to produce toxins. One of these toxins is the real culprit behind the outpouring of water from infected people's small intestines.

Cholera was first described in Sanskrit by Hindu physicians around 400 BCE. The word *cholera*, however, is derived from the Greek term *khole*, meaning "bile." The ancient Greek physician Hippocrates was the first Westerner to mention cholera in his writings.

The transmission of cholera was not well understood until 1854, when John Snow, the father of epidemiology, linked the disease to London's drinking water during a cholera outbreak. Snow hypothesized, correctly, that cholera reproduced in the human body and was spread through contaminated water.

In that same year, an Italian anatomist, Filippo Pacini, was the first to observe with a microscope the comma-shaped bacillus that he isolated from the intestines of its victims. Although Pacini described his findings clearly, Robert Koch is often improperly credited with the discovery of *V. cholera* in 1883.

Although outbreaks and epidemics of cholera have been described for millennia, the disease took its greatest toll in the nineteenth and twentieth centuries, when there were seven pandemics (i.e., worldwide epidemics). In the nineteenth century alone, cholera killed tens of millions of people.

In developed countries, due to nearly universal water treatment and sanitation practices, cholera is no longer a major threat. The last major outbreak of cholera in the United States occurred in 1910–1911.

But cholera is very much still a problem. In October 2010, the worst outbreak in recent years hit less than seven hundred miles off the coast of Florida, in Haiti. By summer of 2017, this outbreak had sickened one million people and taken the lives of ten thousand. Despite an outpouring of humanitarian aid, as well as medical, technological, and public health support—including a cholera vaccination campaign that began in 2016—Haitians continued to fall ill with the disease. And tragically, a major resurgence of the disease occurred in the wake of Hurricane Matthew in October 2016.

In 2016 and 2017, the WHO also sent cholera vaccine to combat large outbreaks that surfaced in the poverty-stricken Horn of Africa—Sudan, Somalia, and Yemen—where, as of mid-July 2017, more than three hundred thousand people had been infected with cholera, and 1,700 had died. By the end of that year, more than a million cases of cholera had occurred in Yemen—the largest cholera outbreak in recent history. At the time, this outbreak was considered by some authorities to be the world's worst humanitarian crisis. Tragically, as of this writing, in early 2019, this cholera epidemic continues in Yemen and is serving as a weapon in an absolutely dreadful war. And in 2019 in Mozambique, cholera vaccinations were administered in the wake of Cyclone Isai.

Preventing cholera seems so simple: provide people with clean drinking water. But this turns out not to be as simple as it sounds. On our planet, about three quarters of a billion people don't have access to safe drinking water. An estimated 2.5 billion people lack access to basic sanitation—that is, functioning toilets and a safe way to treat or dispose of human waste.

A better vaccine against cholera would help, but what would really be transformational would be a vaccine against poverty.

EPIDEMICS OF THE MODERN WORLD

In the twenty-first century, the developing world looks more and more like the developed world. This is both good and bad news.

The good news is that infectious diseases are no longer the major cause of death worldwide. The bad news is that the same chronic diseases that are killing most adults in industrialized countries—heart disease, stroke, chronic respiratory diseases, cancer, and diabetes—are now also, by far, the leading causes of mortality in the developing world. Chronic illness is now responsible for more than 70 percent of all the world's deaths. Furthermore, four out of five deaths from chronic disease now occur in low- and middle-income countries. This epidemic of chronic diseases is an underappreciated contributor to poverty, and a serious hindrance to national economic development.

Exactly why chronic diseases have increased so markedly in recent decades isn't clear. Improved hygiene and vaccine campaigns in developing countries likely played a role in the relative decline of infectious diseases as a cause of death. Martin Blaser has proposed an intriguing hypothesis to explain the rise in chronic diseases—the loss of germs that promote health. As you may recall from chapter 3, elimination of bacteria from the gut microbiome by antibiotics given in childhood is implicated in the development of obesity, type 2 diabetes, allergies, and other chronic diseases. In an article in *Nature Reviews Immunology* in 2017, Blaser theorizes that losses of particular bacterial species of our ancestral microbiota have altered immunological, metabolic, and cognitive development in early life that results in development of chronic diseases in adulthood. [6]

While chronic diseases are now the main cause of premature death in adults, this isn't true for children. Two common infections—diarrhea and pneumonia—remain the leading killers of children worldwide. In children under the age of five, pneumonia is responsible for 1,300,000 deaths each year; diarrhea claims another 700,000.

Evidence suggests that nearly a third of all cases of severe diarrhea among children, and two-thirds of pneumonia deaths among kids are preventable through the use of vaccines. These unnecessary deaths are an ongoing tragedy.

Even though chronic diseases are now the leading cause of death worldwide, new infectious disease epidemics continue to arise. These remind us of our vulnerability to our mortal enemies. Some, like new strains of influenza, have the potential to outwit our best medical defenses—and to set us back to the early years of the twentieth century.

In the remaining chapters of this section of the book on "Mortal Enemies," we'll look more closely at a small number of the more than 140 infectious diseases that have emerged since 1975. Most of these are zoonotic—that is, they are passed from animals to humans, either directly or via insects. But every one of them has a very important lesson for all of us.

7

KILLER VIRUSES

"An inefficient virus kills its host. A clever virus stays with it."—James Lovelock, British scientist

In the previous chapter, you read stories of several historic epidemics that almost brought humanity to its knees. Of the mortal enemies that caused those epidemics, variola major virus was by far the deadliest. Thank goodness the disease it caused, smallpox, has been eradicated.

This chapter tells the stories of two modern-day, or emerging, viruses: HIV and Ebola virus. These mortal enemies continue to have epidemiologists sitting on the edge of their seats.

By James Lovelock's definition, they are both highly inefficient. Untreated, they kill all (HIV) or most (Ebola) of their victims. But they might also be considered clever. Most of the time, HIV lives in its victims for a decade before rearing its head and causing any symptoms of disease—and then, even with highly active antiviral treatment, it can't be eradicated. Furthermore, both viruses are masters at outfoxing our immune system, and then killing us.

HIV/AIDS

"Sex: the thing that takes up the least amount of time and causes the most amount of trouble."—John Barrymore

The HIV/AIDS Pandemic

In 1981, when the first cases of what became known as HIV/AIDS were reported, I had just begun my career as an infectious disease specialist. Very quickly, it became clear that everything about this new disease was astonishing. I've often thought that if the story had appeared earlier in a novel, nobody would have believed it—it was just too far-fetched.

Niels Bohr, the Danish physicist and Nobel laureate, is said to have remarked, "Prediction is very difficult, especially if it's about the future." This was certainly true of HIV/AIDS. In the first year in the United States, all the cases were clustered in California and New York City, and only three of the many predictions that were made at that time turned out to be correct:

1. Based on the risk groups (sexually active gay males, intravenous drug users, and hemophiliac patients who had received blood transfusions), whatever caused the disease was transmitted sexually, or by contact with blood.
2. Because the most common pathogen was an unusual opportunistic fungus, the agent that caused the disease—later called acquired immunodeficiency syndrome (AIDS)—profoundly suppressed cell-mediated immunity.
3. The discovery of its cause would merit a Nobel Prize in Physiology or Medicine. In 2008, two French virologists, Luc Montagnier and Francoise Barre-Sinoussi, shared the Nobel Prize for their discovery in 1983 of human immunodeficiency virus (HIV)—the cause of AIDS.

In 1981, *nobody* predicted that what appeared to be a localized endemic in the United States would become an explosive pandemic, killing an estimated thirty-nine million people by 2013—or that, by that year, 70 percent of the thirty-five million people living with HIV would reside in Africa. Nobody predicted that the virus would severely devastate the gay male and artistic communities in the United States and abroad. And nobody forecast that, over time, women and men would be equally affected, or that the greatest burden would be suffered by society's most vulnerable, marginalized, and stigmatized members.

The Enemy, Its Target, and the Aftermath

The already burgeoning field of molecular virology was responsible for the discovery of HIV, and that discovery led the way for one breakthrough and discovery after another.

You'll recall from chapter 1 that viruses are very simple germs. They consist of nothing more than sets of genes (DNA or RNA) in a protein coat. In the case of HIV, it's RNA. Like all viruses, HIV commandeers the cells that it infects to make copies of itself. But retroviruses are different from most viruses, because they insert their genomes into their hosts' genomes.

As soon as HIV gains entry to a cell, it transcribes its RNA into the cell's DNA—the blueprint for making proteins—using an enzyme it carries called reverse transcriptase. (The usual sequence of events is for DNA to be transcribed into RNA; thus the designation "reverse.")

Even as viruses go, HIV is very simple. It only has nine genes. But because its DNA gets incorporated in the genome of the host cell, it's difficult for the immune system to eliminate it. And HIV's genes can mutate very rapidly, giving rise to variants that are resistant to both the immune system and antiviral drugs.

Where did the HIV retrovirus come from? Genetic research indicates that HIV actually originated in west-central Africa during the late nineteenth or early twentieth century. Like most emerging pathogens, animals were involved. Most of the evidence suggests that HIV arose from the cross-species transmission of other, closely related retroviruses that are found in primates in Africa. (Primates include gorillas, apes, baboons, chimpanzees, monkeys, lemurs, and humans.)

Early in the HIV/AIDS pandemic, it was recognized that there are actually two types of HIV: HIV-1, which is the predominant culprit worldwide, and HIV-2, found mainly in West Africa.

We know of at least four different strains of HIV-1. It appears that two of these originated in 1908 from chimpanzees in southwestern Cameroon; the other two arose from gorillas in the same region.

HIV-2 originated a little later, in monkeys called sooty mangabeys. People eat primate meat in some regions of Africa; it's possible that, through contact with the blood of infected meat (known locally as bushmeat), related retroviruses spilled over into humans.

We know now that the first identified case of HIV-1 infection was in a man in Kinshasa, in the Democratic Republic of Congo, in 1959. It appears that the virus arrived in the United States in New York City from the Caribbean around 1970. How it got there is unknown.

Before antiviral therapy came along in the late 1980s, HIV killed everyone it infected—a distinction shared by only a small handful of human pathogens. How can such a simple creature with only nine genes outwit and kill *Homo sapiens*, with its roughly two thousand genes and the most powerful brain of any species?

The answer is that HIV has the unique capacity—you might say audacity—to target and grow in CD4 lymphocytes, which are masters of immunity (the topic of chapter 4). CD4 lymphocytes are a type of white blood cell. These cells are positioned throughout the body, mainly in lymph nodes. Once HIV gains access to a CD4 lymphocyte, it replicates like crazy—and then kills the cell.

While some people develop a flulike illness at the outset of infection, most HIV-infected people have no symptoms for the first *ten years*. But as more and more CD4 lymphocytes become depleted, a stage is reached where the immune system begins to be compromised.

When a human being's CD4 lymphocyte count falls below 300 cells/ mm^3 of blood (a normal count is between 500 to 1,500 cells/mm^3), a state of severe immunodeficiency occurs. Then all kinds of opportunistic pathogens can move in, including fungi (such as *Pneumocystis jirovecii*, a leading pathogen in the early years of the AIDS epidemic, and *Cryptococcus neoformans*, currently the leading cause of death among Africans with HIV), bacteria (such as *Mycobacterium tuberculosis*, the cause of tuberculosis), viruses (mainly members of the herpes group), and parasites (such as *Toxoplasma gondii*, formerly the most common cause of brain masses in AIDS patients). Malignancies, such as Kaposi's sarcoma and lymphomas, can also emerge when severe immune deficiency develops.

HIV itself can cause a neurodegenerative form of dementia, something like Alzheimer's disease, in which patients lose their memory, become mute, and stare off into space. I remember with profound sadness caring for patients in this tragic phase of their illness during the early years of the HIV/AIDS epidemic.

Treatment and Prevention

For half a dozen years after the first HIV cases appeared in the United States, almost everyone who contracted the disease died.[1] But in 1987 this profoundly tragic situation began to break when the first antiviral drug, zidovudine (Retrovir), was approved by the Food and Drug Administration (FDA). As its name suggests, Retrovir inhibits the activity of HIV's reverse transcriptase enzyme. Soon afterward, several pharmaceutical companies—to their great credit—developed many other drugs with similar effects. And the race was on for pharmaceutical companies to develop better and better antiretroviral agents.

The real game changer came in 1996, when multiple studies revealed that combinations of antiviral drugs, called HAART (highly active antiretroviral therapy), provided dramatic benefit. When treated with HAART, AIDS patients literally got up from their deathbeds and resumed normal lives. By this time, treatment of HIV/AIDS had become so complicated that many infectious diseases physicians specialized only in their management. In the mid-1990s, the joy of my colleagues who had dedicated their lives to caring for these patients was palpable.

Today, there are at least twenty-seven medicines approved by the FDA to treat HIV infection. When three or four of these drugs are combined, they elicit a marked reduction in the amount of HIV in the blood (called the viral load). This antiviral effect is associated with a significant increase in the CD4 lymphocyte count and the restoration of immunity.

That's the good news. The bad news is that HAART doesn't completely eradicate the HIV virus. The virus goes into hiding, so treatment must be lifelong.

Still, HAART treatment has been so successful that in many parts of the world, HIV has become a chronic rather than a fatal condition. Today, the main causes of death of successfully treated patients are the same as those of noninfected people: cardiovascular disease and cancer.

In the early years of HAART, many studies were carried out to determine when treatment should be started. The studies made it clear that treatment should begin as soon as HIV infection is diagnosed.

In recent years, single daily doses of three or four drugs, combined in one pill, became available. New studies suggest that a single injection of a long-acting two-drug regimen, given every eight weeks, works as well as the three daily pills. If these results are confirmed by other studies, this could further transform HIV treatment.

Another remarkable outcome of the HIV/AIDS pandemic has been the worldwide response to this health disaster. To prevent emergence of resistance of HIV, all patients need to be on a combination of antiretroviral drugs. Because such combinations are very expensive, treatment at first was feasible only in developed countries. But, amazingly, due to the leadership of many heroes in medicine, public health, the pharmaceutical industry, government, and nongovernmental organizations—as well as celebrities and patient advocacy groups—in recent years this inequity has greatly diminished. Built upon growing evidence of the lifesaving capacity of combination drug treatment, two extraordinary partnerships were born: the Global Fund to Fight AIDS, Tuberculosis and Malaria in 2002 and the U.S. President's Emergency Plan for AIDS Relief (PEPFAR) in 2003.

Launched out of humanitarian concern by president George W. Bush—with unprecedented bipartisan support—PEPFAR has supported antiviral treatment throughout poverty-stricken countries in Africa. Because of funding provided by PEPFAR and nonprofit nongovernmental organizations, by mid-2016 18.2 million people living with HIV in Africa were receiving combination antiretroviral therapy. As a result, AIDS-related deaths fell by almost 50 percent from 2005. In 2013, John Kerry announced that the one-millionth baby had been born HIV-free because of treatment of HIV-positive pregnant women—all made possible by PEPFAR.

Strategies for HIV prevention have also gained ground. Early on, the development of a screening test for HIV eliminated contaminated blood products from the blood supply. Mother-to-child transmission of HIV has been all but eliminated by treating infected mothers.

Recent studies suggest that, when taken daily, a pill containing a combination of antiviral drugs significantly reduces the risk of HIV infection for uninfected men who have sex with men. Called pre-exposure prophylaxis, or PrEP, this approach to prevention is now recommended for all HIV-negative men at high risk of acquiring infection when having sex with men. (While PrEP has many people cheering, concern also has been raised that many men may now forgo using condoms to prevent infection. More troubling still, very few men are actually using PrEP. At the 2018 Conference on Retroviruses and Opportunistic Infections in Boston—

nearly six years after the FDA's approval of PrEP—it was reported that only a small fraction of men who can benefit from the drug are using it.)

The HIV/AIDS epidemic could be brought to an end if only all infected men and women would stop having unsafe sex, and all intravenous drug users would stop sharing needles. Both of these changes in behavior are of course much easier said than done.

Lessons for the Future

"The nature of a protective immune response to HIV is still unclear. Because in a very, very unique manner, unlike virtually any other microbe with which we're familiar, the HIV virus has evolved in a way that the immune system finds it very difficult, if not impossible, to deal with the virus."—Anthony Fauci, director, National Institute of Allergy and Infectious Diseases

Despite extraordinary advances in therapy, 54 percent of the 36.9 million people living with HIV infection in 2014 didn't receive treatment. In the same year, there were two million new HIV infections—fifty thousand of them in the United States. And because the infection is usually silent and invisible for ten years, most HIV-infected people don't even know they are infected. Nonetheless, in July 2017 the UN AIDS Agency announced that for the first time in the global AIDS epidemic more than half of all those infected with HIV were on antiviral therapy.

Successful treatment regimens have led to complacency and a belief that HIV/AIDS is under control. But for the many millions of individuals living with HIV, this isn't the case. They must commit to lifelong treatment with expensive medications and to sustained vigilance regarding protected sex. The good news, however, is that the pharmaceutical industry is working on new drugs that can eradicate the latent reservoir of cells that harbor HIV DNA in their genomes.

Despite the daunting challenges, leaders at the National Institutes of Health, the Centers for Disease Control and Prevention, and the World Health Organization are optimistic. They suggest that a universal rollout of combination antiretroviral therapy could halt the pandemic by 2030. The unexpected announcement by President Trump of his desire to end HIV/AIDS, in his 2019 State of the Union address, set the stage for achieving this goal in America.[2] On the other hand, the emergence of highly drug-resistant strains of HIV in the developing world, and concerns about the continued funding of PEPFAR, may prevent us from reaching this goal.

As you would expect, development of an HIV vaccine has been a top priority since the discovery of the virus in 1983. This goal is shared by many of the world's most brilliant virologists and immunologists. But because of HIV's unique capacity to incorporate itself into the genome of CD4 lymphocytes—a pivotal cell in human immunity—and to evolve mechanisms of resistance to the immune system, none of the one hundred different vaccines that have been tested proved successful. However, in 2015 a unique vaccine developed by Robert Gallo, a pioneer HIV/AIDS scientist, entered clinical trials in the United States.[3] And in 2016 a new HIV vaccine trial was initiated in Johannesburg, South Africa. These trials offer hope that the goal of eradication of HIV may someday be achieved, as it was for smallpox in 1976.

EBOLA

> "Riots are breaking out. Isolation centres are overwhelmed. Health workers on the frontline are becoming infected and are dying in shocking numbers."—Joanne Liu, international president, Médecins Sans Frontières

Ebola Epidemics

Of the more than 140 current emerging infections, Ebola virus disease (also referred to as Ebola hemorrhagic fever, or simply Ebola) has an astonishing capacity to strike fear and panic into human hearts.

Like HIV, Ebola virus got its start in Africa. It too appears to have spilled over into humans from contact with infected animals. But unlike HIV, which is insidious but controllable, Ebola virus claims the lives of its victims (over 50 percent of them) in a rapid and gruesome manner.

After an incubation period from four to nine days following initial exposure, the illness usually begins with the sudden onset of fever and chills, followed by flulike symptoms (muscle pain, runny nose, and cough), gastrointestinal symptoms (diarrhea, nausea, vomiting, and abdominal pain), and, in the most severe cases, internal and external bleeding (from the eyes, ears, and mouth). In the terminal stage of illness (days 7 to 10), confusion sets in, and the victim falls into a coma. Shock develops from the dehydration (caused by diarrhea and vomiting) and bleeding.

Ebola virus is a master of evading the body's immune system. In fact, it will actually infect immune cells, using them to travel through the body to the liver, kidneys, spleen, and brain.

Ebola was first recognized in 1976 when two epidemics occurred simultaneously in Zaire (currently the Democratic Republic of Congo, or DRC) and Sudan. Since then, twenty-one outbreaks have occurred, mostly in countries in Equatorial Africa.

The epidemic that began in December 2013 was by far the largest and most frightening. For the first time, countries in West Africa—mainly Guinea, Liberia, and Sierra Leone—were hit. The strain of the Ebola virus that was first isolated in Zaire (dubbed EBOV) is the culprit. How it found its way to West Africa is unknown—though scientists suspect fruit bats as the carrier. Experts think fruit bats carried Ebola from Equatorial Africa to West Africa, and it then jumped to humans through close contact with the blood, secretions, organs, or other bodily fluids of infected animals—fruit bats, chimpanzees, gorillas, monkeys, forest antelope, and porcupines. West Africans eat many of these animals, so it's possible that the disease spread to humans through hunting and meal preparation.

The epidemic began in a one-year-old boy in Guinea, then spread to Liberia and Sierra Leone. Small outbreaks and isolated cases occurred in other nearby African countries. Outside of Africa, imported cases were seen in the United Kingdom and Sardinia. In Spain and the United States, imported cases led to secondary infections of medical workers. This caused much panic—but, fortunately, the disease didn't spread farther.

Ebola infection occurs mainly through contact with the skin or body fluids of an infected person. Thus, most cases occur among people who provide direct care to Ebola patients—usually family members and healthcare professionals. (About one-quarter of cases occur among healthcare workers.) Traditional funerals in which family members prepare infected corpses for burial also transmit the illness to household members.

Many of the people who survive Ebola have a slow and painful recovery that includes fatigue, loss of appetite, hair loss, and eye diseases. The virus can persist in breast milk and semen for many months after the cessation of symptoms. (Recent studies detected the Ebola virus in the semen of 1 to 2 percent of male survivors a full year after recovery—and in the semen of one man 565 days after he became infected.) During this period, it can be transmitted through sex and through the nursing of infants.

Although there is some concern that Ebola might mutate and become transmissible through the air, fortunately that hasn't happened.

As of April 2016, more than two years after the first case in Guinea, a total of 28,616 cases and 11,325 deaths were recorded. The World Health Organization believes that the actual number of deaths was much larger. A recent estimate of the cost of West Africa's Ebola epidemic, which combines the direct economic burden and the indirect social impact, is $53 billion—truly astounding.

Liberia, Guinea, and Sierra Leone are all poor nations whose health-care systems were unprepared for, and overwhelmed by, the devastating illness. Members of afflicted families and villagers were terror stricken; healthcare providers were shaken; dead bodies lay in the streets, sometimes for days. At times, the havoc caused by Ebola was reminiscent of the Great Plague of London 350 years earlier.

The Enemy, Its Targets, and the Aftermath

Ebola is caused by four of five species belonging to the *Ebolavirus* genus in the Filoviridae family, all of which are RNA viruses. The name *filoviridae* comes from the Latin *filo*, meaning "threadlike." Not much is known about filoviruses, since their deadliness and their highly infectious nature make them difficult to study. Here, however, is what we do know:

The first two of these five viruses were discovered in 1976 by a team of scientists from the Center for Infectious Diseases and Prevention, along with Peter Piot, a Belgian microbiologist who is now the Director of the London School of Hygiene and Tropical Medicine. Upon examining a blood sample taken from a Belgian nun working in Zaire, they were surprised to see a "gigantic worm-like structure—gigantic by viral standards" revealed by an electron microscope. They named the virus after the Ebola River, which runs close to the village of Yambuku, in Zaire. Even at that early juncture, the research team was able to accurately describe much about Ebola virus disease. They emphasized the critical importance of the safe handling of corpses and the value of quarantining infected people, in order to control epidemics.

In 2015, researchers sequenced ninety-nine Ebola virus genomes from seventy-eight patients. They found 341 different genetic changes that made the strain behind the 2013–2015 outbreak distinct from previous outbreak strains. (Here's how dangerous Ebola is: five members of the research team, all of whom took maximum precautions, became ill and died from Ebola before the study was published.)

Treatment and Prevention

As a rule, treatment of any disease needs to be started as soon as possible—preferably before symptoms develop (as with HIV infection). But many of the early symptoms of Ebola—fever, chills, respiratory problems, and gastrointestinal symptoms—are common to many other illnesses. This makes establishing an early diagnosis challenging, especially in developing countries—as well as in locations far removed from an Ebola epidemic.

In 2014, a Liberian visitor to the United States who had Ebola was seen in a Texas emergency room. He was sent home because he didn't have a fever. It wasn't clear at the time that 18 perecnt of Ebola patients don't come down with fevers. Meanwhile, in Minnesota, which has a large Liberian immigrant population, when visitors from Liberia showed up in emergency rooms with fevers in 2014–2015, panic over Ebola sometimes ensued. (While my infectious disease colleagues and I were on high alert, no cases of Ebola actually reached Minnesota. The sick visitors from Liberia had malaria and other easily treated diseases.)

At present, there is no cure for Ebola. The treatment of Ebola virus disease is largely aimed at reducing symptoms and supporting vital bodily functions. Diarrhea can be prolific, contributing to dehydration and the collapse of veins, so resuscitation with fluids and the monitoring of blood electrolytes are of paramount importance.

During the West African Ebola epidemic, several experimental drugs and immunological treatments were used on a case-by-case basis. Some of these agents looked particularly encouraging when tested on monkeys. One drug, ZMapp (a combination of three antibodies that target Ebola virus), showed evidence of reduced mortality in a small randomized clinical trial. But because the epidemic began to wane just as these new treatments came online, an insufficient number of patients was available for large controlled trials. It was suggested that if a resurgence of the epidemic occurs, or a new Ebola outbreak appears, the most promising treatments would be properly tested. (Indeed in 2018, when the tenth outbreak of Ebola occurred in the DRC, clinical trials of two monoclonal antibodies were launched. And by 2019, both had shown very promising early results.)

Many of the lessons for preventing the spread of Ebola that were learned forty years earlier, in the first Ebola epidemics, were relearned during the West Africa epidemic. It's essential to follow established guidelines for quarantining patients—*and* the people they come into contact with, including exposed travelers and healthcare professionals arriving from affected countries. Hospital workers caring for patients must wear special protective gear, and extraordinary care needs to be used in putting on, taking off, and disinfecting or destroying the gear after any patient contact.

The most exciting news for the prevention of Ebola came from the results of a clinical trial in Guinea in 2015, in which a tested single-dose vaccine prevented Ebola infection in 100 percent of the people immunized. Although more research is needed, former WHO director-general Margaret Chan called this vaccine "an extremely promising development . . . for both the current and future Ebola outbreaks." In 2017, the results of several successful phase 1 clinical trials of Ebola vaccines (aimed mainly at determining their safety) were reported. So, if (some said *when*) Ebola rears its ugly head once again, it looks like we're much better prepared.

Indeed, when a small Ebola outbreak in the Democratic Republic of Congo emerged in April 2017, the vaccine, made by Merck and stored in the United States, was ready to go. (Ebola virus is considered endemic in the DRC, where ten outbreaks have been recorded since 1976.) Fortunately, by July 2017, that outbreak, which claimed the lives of four people, was declared over by the WHO, and vaccination wasn't needed.

But, sure enough, in July 2018 Ebola resurfaced in the DRC. To complicate matters further, this Ebola outbreak is the first to occur in a war zone. By mid-December, 505 Ebola cases were confirmed, with a death toll of 298. And by the end of April 2019, this new outbreak had sickened close to 1,400 people and killed 900. Particularly disheartening is that this rapidly evolving outbreak occurred despite the existence of a powerful vaccine.[4]

The good news, however, is that, unlike the sluggish response to the Ebola epidemic in West Africa, international health agencies are rapidly scaling up the expertise and resources to combat this outbreak. Investigational drugs, such as monoclonal antibodies, including ZMapp, are being administered; new individual air-conditioned, biosecure cubicles are being used in the care of patients; and a vaccination program is underway.

Lessons for the Future

> "Perhaps the only good news from the tragic Ebola epidemic in Guinea, Sierra Leone, and Liberia is that it may serve as a wake-up call: we must prepare for future epidemics of diseases that may spread more effectively than Ebola."—Bill Gates

In August 2014, five months after the first cases of Ebola were reported in Guinea and Liberia, the WHO declared the outbreak a public health emergency of international concern. By then, the virus had killed thousands of patients and healthcare workers in Guinea, Liberia, and Sierra Leone, and full-scale panic had set in. Considering how understaffed and underfunded those countries' health services were, ending the epidemic in two years was a major public health achievement. The public health, humanitarian, and scientific response was huge. It involved many governmental and nongovernmental organizations, including the WHO, the CDC, Médecins Sans Frontières/Doctors Without Borders, the National Institutes of Health, the Bill & Melinda Gates Foundation, and others.

Nevertheless, had the world cared enough—and paid sufficient attention—the epidemic might have been halted much earlier, and thousands of lives might have been saved. I strongly suspect that, had the epidemic begun in the developed world rather than the developing world, efforts would have ramped up far more quickly. The WHO in particular was harshly criticized for failing to provide an effective oversight role.

Reassuringly, by the end of 2015, a plethora of expert panels had considered what was done well and what was done badly. With the gift of hindsight, the shortcomings of the WHO and other organizations were identified, and recommendations were made for coordinated stewardship in responding to future infectious disease outbreaks. In May 2016, the recommendations of four global commissions on Ebola were published in the journal *PLOS Medicine*. All the experts agreed that there exists a worldwide moral obligation to ensure that the same mistakes aren't repeated.

No sooner than the WHO declared the Ebola epidemic in West Africa over on January 14, 2016, a twenty-two-year-old student in Sierra Leone tested positive for the disease. This case, and the recent epidemics in the DRC, underscore the importance of sustained vigilance in both West and East Africa. As everyone in the field knows, the virus remains hidden in animals, usually bats, and is almost certain to spill over into humans again.

In developed countries such as the United States, better communication networks were created among public health officials, hospitals, and care providers. Procedures for infection control were tightened, to help keep citizens and healthcare professionals safe during future epidemics.

By the end of 2015, only four cases of Ebola virus disease were identified in the United States. The first was the Liberian visitor to Dallas, who died. The second was a doctor from Guinea, who was treated successfully in New York City. The other two cases were healthcare providers who had contact with the patient in Texas. Both survived.

For those of us who lived through the early days of the AIDS epidemic in the United States, the brief national panic over Ebola brought back some very bad memories. Panic, paranoia, and discrimination were rampant then as well. Sensationalism often overpowered responsible journalism. But fortunately, wisdom and compassion prevailed in both cases—and with time, science-based public health measures eventually turned the tide of both epidemics.

As I write these words in August 2019, many of the lessons learned from the Ebola outbreak in West Africa are being applied in fighting the current outbreak in the DRC. Large-scale use of the Ebola vaccine and preliminary results of treatment trials with monoclonal antibodies are highly encouraging. Nevertheless, it is too soon to tell when this disease will peter out. On July 17, 2019, the WHO declared that the Ebola outbreak is an international emergency. Dr. Vinh-Kim Nguyen, the medical team leader for Médecins Sans Frontières (Doctors Without Borders), has underscored the importance of listening to and gaining the trust of the community in this battle. In the long run, if Ebola is to be eradicated, it will also be essential to acknowledge and address issues such as poverty and injustice.[5]

8

THE BUZZ ON MOSQUITO-BORNE INFECTIONS

"If a mosquito has a soul, it is mostly evil. So I don't have too many qualms about putting a mosquito out of its misery. I'm a little more respectful of ants."—Douglas Hofstadter, Distinguished Professor of Cognitive Science, Indiana University

Mosquitoes are the single most deadly animal on Earth, killing millions of humans and other animals every year. They do so by serving as transmitters of pathogens—most notably, *Plasmodium*, the parasite that causes malaria, and a group of viruses called arboviruses.

You'll recall from chapter 6 that, of the 3,500 species of mosquitoes worldwide, *Plasmodium* is carried by only about forty members of the genus *Anopheles*. The three arboviruses discussed in this chapter (dengue, chikungunya, and Zika) are carried by other mosquito species, which belong to the genus *Aedes*.

All mosquitoes have two things in common, however. First, only the females bite and draw blood, which contains the proteins needed to make their eggs. And all mosquitoes are dependent on both water (where they lay their eggs) and warm temperatures.

This is why more species are found in tropical countries. Brazil has 450 mosquito species, while 166 species are found in the continental United States, and Norway is home to a paltry sixteen species. But the number of species doesn't necessarily equate with the total number of actual insects, as anyone who lives in Minnesota—where we joke that the mosquito is the state bird—can attest.

Mosquitoes have an amazing sense of smell, which is how they home in on their prey. They are attracted to carbon dioxide (released from our lungs) and chemical compounds emitted from our skin.[1] Recent studies suggest that the composition of the one trillion bacteria making up your skin microbiome, which you read about in chapter 3, determines whether your particular odor is attractive to mosquitoes. If it is, then you are among the 20 percent of people who are high attractors.

Here are some more amazing mosquito facts: they are found in every country of the world except Iceland and Antarctica; they appeared on Earth at least two hundred million years before our own species, *Homo sapiens*; and they are an essential food source for many birds, bats, and fish. In fact, without mosquitoes, many of the world's ecosystems would collapse. So next time you angrily swat one of these pests, think about all the other animals that can't live without them.

DENGUE

> "We've seen from Ebola that, in this global world that we're living in . . . infectious diseases can travel around; the conditions for these diseases are dynamic over time and, given that we're changing our social and environmental dynamics, the global distribution of these infectious diseases, like dengue, is going to change."—Corrine Schuster-Wallace, program officer, United Nations University Institute for Water, Environment and Health

The Dengue Pandemic

Dengue has been around for a very long time. The first case of what was probably dengue is recorded in a Chinese medical encyclopedia from the Jin Dynasty (265–420).

In our area of the world, the first confirmed case was reported by Benjamin Rush, who in 1789 coined the term "break-bone fever," referring to the severe muscle, bone, and joint pains associated with dengue.

The four types of dengue virus, DENV 1, DENV 2, DENV 3, and DENV 4, are members of the *Flavivirus* genus. Other flaviviruses include Zika virus, and the nastiest of them all, yellow fever virus, from which their name is derived (*flavus* means "yellow" in Latin). (Currently, central Africa is witnessing a serious resurgence of yellow fever, and in 2017 Brazil ordered a massive supply of yellow fever vaccine in response to a major upsurge of cases.)

All four forms of the dengue virus are carried by two mosquito species, *Aedes aegypti* and *Aedes albopictus* (also known as the Asian tiger mosquito). Today about 2.5 billion people, 35 percent of the world's population, live in areas of the world where these mosquito species hang out, thereby setting the stage for the explosive dengue pandemic that started in the 1970s. (Thomas Frieden, director of the Centers for Disease Control and Prevention, recently called *Aedes aegypti* "the cockroach of mosquitos" because it tends to live in and around human homes. On top of that, it prefers human blood, which has been shown to increase the virus's reproduction rate. And it is a very sneaky biter, attacking subtle places such as the ankles.)

As of 2018, dengue was a problem in at least 120 countries in Asia, the Pacific, Africa, the Americas, and the Caribbean. Sri Lanka was hit particularly hard, with over one hundred thousand cases and 296 deaths by mid-2017. Vietnam reported about the same number of cases—a 42 percent increase over the same period in 2016. And in 2019, Central America was grappling with its worst outbreak of dengue fever in decades.

The World Health Organization estimates that fifty to one hundred million infections—including half a million cases of the most severe form of the disease (dengue hemorrhagic fever)—occur annually, resulting in twenty-two thousand deaths each year. Some experts estimate numbers more than three times those of the WHO. Half a million people are hospitalized each year with dengue. And in the United States alone, dengue illness costs, on average, $2.1 billion a year.

What happened in the past fifty years that explains the extraordinary spread of dengue? One main cause is unplanned urbanization, leading to inadequate water, sewage, and waste management. Another is increased international travel to affected areas. Nearly all cases reported in the continental United States, where contact with *Aedes* mosquitoes is infrequent, involved travelers or immigrants who caught the disease elsewhere. (However, small outbreaks of indigenous dengue have been reported in Florida and Texas.)

During my sixteen years as the director of the University of Minnesota Medical School's International Medical Education and Research Program, our office counseled well over five hundred medical students who traveled to tropical countries to gain clinical experience. While mosquito-borne pathogens, such as *Plasmodium falciparum* and dengue viruses, were officially the leading infectious disease threat to their lives, in fact their biggest risk of dying—by far—was getting into a motor vehicle accident.

The Enemy, Its Targets, and the Aftermath

Exactly where dengue came from is unclear, but it appears that the four dengue viruses originated in monkeys and independently jumped to humans in Africa or Southeast Asia between one hundred and eight hundred years ago.

In 1907, two young officers of the U.S. Army Medical Corps, P. M. Ashburn and Charles F. Craig, were sent to the Philippines to study dengue. They were the first to demonstrate that the disease was caused by a virus. It wasn't until 1943, however, that a dengue virus (DENV 1) was isolated by Ren Kimura and Susumu Hotta. Some years later, the other three types were identified as well.

The dengue virus genome is a single strand of RNA. The four types are closely related—they share approximately 65 percent of their genomes—but each has its own unique interactions with the antibodies in human blood. The four forms of dengue cause the same symptoms—but it's possible to be immune to one form but vulnerable to the other three (or immune to two and vulnerable to the other two, etc.).

After an incubation period of four to eight days, infection by any of the dengue viruses results in a wide spectrum of symptoms. The good news is that, in 80 percent of cases, infected people experience no symptoms at all. If symptoms do develop, they include sudden high fever (temperatures up to 106 degrees F); headaches; pain behind the eyes; severe muscle, bone, and joint pain; and nausea and vomiting. In many cases, a skin rash appears two to five days after the onset of fever.

Fortunately, the great majority of patients recover within two to seven days. But as many as 5 percent of all dengue patients develop severe, life-threatening dengue hemorrhagic fever. In this unfortunate minority, the disease proceeds to a critical phase just as the fever is resolving. During this period, there is leakage of plasma from blood vessels, which can accumulate in the chest and abdominal cavity. This depletes the amount of fluid circulating in the body, often resulting in shock and a decreased blood supply to vital organs. Severe bleeding from the gastrointestinal tract also typically occurs.

Prior infection with one type of dengue virus not only doesn't provide immunity against the other three—it makes you more vulnerable to them. If you've already had one type of dengue and you become infected with another, the earlier infection *increases* your risk of developing severe symptoms.

In recent years, improved diagnostic tests and guidelines for recognizing the warning signs of severe disease have helped us fight the virus. Thus, patients with severe dengue can be quickly diagnosed and moved to intensive care units. But an intensive care unit—and someone to make a quick diagnosis—aren't always readily available. Because most cases of dengue occur in poor countries, over twenty thousand people die from the illness annually. Sadly, most of these fatalities are in children.

Treatment and Prevention

To date, the U.S. Food and Drug Administration hasn't approved any drugs against dengue, but substantial efforts are underway to develop them. Current treatment focuses on alleviating symptoms—and, for those with severe disease, on administering blood transfusions or otherwise replacing body fluids.

Remarkable headway is also being made in developing vaccines. In Latin America, a vaccine that targets all four dengue viruses has been tested—and has produced encouraging results. This vaccine, called Dengvaxia, was developed by the French pharmaceutical giant Sanofi Pasteur. The vaccine is licensed in twenty countries, though in early 2019 it was actually available only in ten.

In its current form, though, this vaccine has some drawbacks. On the one hand, it reduced the risk of adults developing dengue by about 60 percent; on the other, it wasn't as effective in children, and it may even have increased the risk for kids under the age of six. (And, tragically, use of the Sanofi vaccine was dramatically curtailed in 2017, when fatal cases of dengue were reported in people who had never contracted the illness before they were vaccinated.)[2]

Another vaccine trial that has everyone in the field buzzing was carried out in 2015. In this trial, twenty-one healthy adult volunteers were each given a single dose of a live but weakened vaccine, which was active against all four types of dengue. When the volunteers were later challenged with live dengue viruses, 100 percent were protected—apparently against all four forms of the virus. Meanwhile, the twenty volunteers who received a placebo vaccine all got sick. In the words of Duane Gubler, a veteran in the field of dengue research, "For the first time in 50 years, I'm really enthusiastic and confident that we will have a vaccine, and we'll be able to use the vaccine in the next few years." This research also has positive implications for the development of vaccines against other arboviruses, such as Zika.

Innovative approaches to controlling mosquitoes may also help to limit the spread of the dengue viruses. One strategy uses a symbiotic bacterium, *Wolbachia*, adapted from the fruit fly, which is sprayed in areas where the *Aedes* mosquitoes live. The bacterium both shortens the life of the mosquito *and* blocks transmission of the dengue virus by a mechanism that isn't completely understood.

A biotechnology start-up, MosquitoMate, recently applied for and received approval from the U.S. Environmental Protection Agency to use *Wolbachia* as a pesticide. And according to a report in *Nature News* in August 2017, a project using *Wolbachia*-infected mosquitoes has the potential to eliminate mosquitoes from a number of islands in the South Pacific in ten years. A trial of *Wolbachia*-infected mosquitoes in Townsville, Australia (population 187,000), was reported to have dramatically reduced the rate of dengue in 2018. (*Wolbachia* is arguably the most successful bacterium in the world. It infects at least 40 percent of all arthropod species—insects, spiders, scorpions, etc. And it is a master manipulator of the sex life of its host; it both feminizes and kills off males.)

But until a highly effective vaccine is available, here's what you need to know when traveling to any of the more than one hundred countries where dengue is a problem:

First, bring mosquito repellent—one containing 20 to 30 percent DEET. Lemon eucalyptus oil or picardin also work.

Second, wear long-sleeved shirts and pants—especially during the day, when *Aedes* mosquitoes feed.

Third, use screens in all windows and doorways, and quickly repair any holes.

Fourth, and most importantly, remove *all* containers that can hold water (even bottle caps), because mosquitoes might breed in them. Keep glasses, cups, saucers, ashtrays, and the like inside closed cupboards or cabinets.

Pyrethrin-impregnated bed nets are very effective in controlling *Anopholes* mosquitoes. But these mosquitoes feed at night—and they're not the type of mosquito that transmits the dengue virus. Bed nets aren't that valuable in fighting *Aedes* mosquitoes—the dengue carriers—because *Aedes* mosquitoes bite during the day. But using such netting over infant carriers when you're outside during the daytime is a good idea.

Lessons for the Future

"Infectious diseases—most of which are preventable—disrupt the lives of millions of Americans each year. But the country does not sufficiently invest in basic protection that could help avoid significant numbers of outbreaks and save billions of dollars in unnecessary healthcare costs."—Trust for America's Health Infectious Diseases Policy Report (2015)

Due to the geography that favors *Aedes* mosquitoes, the greatest risk of mosquito-borne infections—including dengue, chikungunya, and Zika— is borne by those who can least afford to treat them. While Americans and Europeans have become increasingly complacent about the risk of infectious diseases, for people living in poverty-stricken countries, these infections are just a mosquito bite away.

Many dedicated researchers, pharmaceutical companies, governments, and nonprofits have done a great deal to reduce the impact of mosquito-borne infections. But additional funding for these efforts is sorely needed. And because climate change is at least partially responsible for where mosquitoes—and the diseases they carry—end up, the risk of *Aedes* making its way to northern climes is a real concern. (More about this topic in chapter 20.)

CHIKUNGUNYA

"The tiger mosquito offered chikungunya what amounted to frequent flier miles on a fleet of jets bound for cooler climes. Within a few years, the virus showed up in Italy and France, ferried from person to person by black-and-white striped tiger mosquitoes."—Nathan Seppa, writer, *Science News*

The Chikungunya Pandemic

Chikungunya got its name from a word in the Makonde language of East Africa that means "to walk bent over." This refers to the severe joint pain that is a hallmark of the disease.

Chikungunya virus, the mosquito-borne virus that causes the disease, was first isolated in 1952 in present-day Tanzania. Historically, the virus circulated mainly in Africa, with periodic brief outbreaks in other areas of the world documented as early as the eighteenth century.

The reason chikungunya is considered an emerging infection is because of the astoundingly swift spread of the virus outside its usual geographical boundaries in recent years. Epidemics have appeared in countries in the Indian Ocean, the Pacific Islands, and, in 2013–2014, the Americas. Today an estimated three million infections occur each year. While the mortality rate is low (less than 1 in 1,000), the severe chronic joint pain is both common and very disabling.

In 2005, an outbreak of chikungunya occurred for the first time on Reunion Island, a French territory in the Indian Ocean. An estimated 266,000 people (35 percent of the population) were affected. This seemed an unusual location for an outbreak because Reunion Island had few or no *Aedes aegypti*, the mosquito species usually favored by chikungunya virus. Researchers soon figured out that the African strain of chikungunya that hit the island had mutated to thrive inside a different mosquito, the Asian tiger mosquito, or *Aedes albopictus*. This mosquito is not only more aggressive, but it can also live outside of tropical climes.

Within a few years of the Reunion outbreak, the tiger mosquito showed up in temperate areas of Italy and France, and hundreds of Europeans were infected with chikungunya. And in 2013, chikungunya surprised everyone when it showed up for the first time on the island of St. Martin in the Caribbean. Within a year and a half, chikungunya established a foothold throughout the Caribbean, as well as in Central America, the northern parts of South America, and even Florida (where eleven people had contracted the illness by 2015). In 2016, it was detected for the first time in mosquitoes in Turkey.

As with dengue, unplanned urbanization and global travel were key factors in the spread of chikungunya.

The Enemy, Its Targets, and the Aftermath

Chikungunya virus was identified by R. W. Ross in 1956, following the 1952 outbreak of the disease in Tanzania. Like dengue virus, the chikungunya virus genome is made up of a single strand of RNA.

The great majority of people who become infected with chikungunya virus become ill. The illness resembles dengue in many ways: symptoms last five to seven days and typically include high fever, headaches, and exhaustion. However, a distinguishing feature is the crippling joint pain in the legs, arms, hands, and feet. This pain can last months, or even years. There also appears to be a link between chikungunya and an increased risk of serious brain inflammation, known as encephalitis. Unlike dengue, however, chikungunya is rarely life-threatening, and it doesn't cause bleeding.

Patients with severe chikungunya should be hospitalized. Those with the greatest risk of dying are newborns, the elderly, and people with underlying medical conditions.

Pregnant women face special risks. Of thirty-nine pregnant women on Reunion Island who had chikungunya around the time of labor, nineteen delivered infected babies. Ten of these infected newborns suffered serious complications, including swelling of their brains and developmental abnormalities.

Chikungunya virus is transmitted from person to person, not just by mosquito bites. It's also possible to contract the virus through contact with infected monkeys, birds, cattle, and rodents, all of which can harbor the virus.

Treatment and Prevention

Anti-inflammatory drugs can help control symptoms such as joint pain. Otherwise, there are currently no treatments or vaccines. However, potential vaccines are in the works. Promising results of a phase 2 clinical trial of one such vaccine, MV-CHIK, were published in 2018.[3] Other vaccines are in the early stage of development.

In the meantime, strategies for prevention are similar to those for dengue outlined above.

Lessons for the Future

"Make everything as simple as possible, but not simpler."—Albert Einstein

There are no simple solutions to eradicating arboviruses such as chikungunya. They have millions, if not billions, of years of evolutionary history on their side.

The good news is that we are getting more knowledgeable about how to develop an effective vaccine and how to control the mosquitoes that spread the illnesses. If *Aedes albopictus* sparks new outbreaks in temperate regions of the United States and Europe, increased urgency to find a vaccine will follow.

ZIKA

"If you think you are too small to make a difference, try sleeping with a mosquito."—the Dalai Lama

The Zika Pandemic

Just about when I learned how to pronounce and spell *chikungunya*, another mosquito-borne disease began terrorizing the Americas: Zika.

Zika shares many of the same features as dengue and chikungunya. But the Zika pandemic is different in two important ways:

First, the spread of Zika has been truly explosive. Brazil was the first country in the Americas to report Zika, in May 2015. In less than a year, over one and a half million cases were recorded in that country. Within months, the pandemic had spread to thirty-three other countries or territories in the Americas. On February 1, 2016, the World Health Organization declared Zika a "public health emergency of international concern."

Second—and the main reason Zika is so frightening—the Zika virus targets the nervous system. In particular, it can damage infants in the womb, so that they are born with very small heads, brain damage, or both. The images, now seen all over the world, of mothers holding their newborn babies with little heads are heart wrenching.

Zika virus was discovered in 1947 by researchers studying yellow fever in the Zika Forest of Uganda. The first case was in a caged macaque monkey. Until 2007, however, scientists knew of only fourteen cases of the disease in humans. That year it arrived on the Southwestern Pacific island of Yap. Within a few months, nearly three-quarters of the island's residents above the age of three had been infected. On Yap, the illness was generally mild, and no one died.

Then, in 2013, Zika popped up in Tahiti and other parts of French Polynesia. An estimated twenty-eight thousand people (just over 10 percent of the population) were sick enough to seek medical care.

After circulating around Polynesia, Zika arrived in the Americas—on Chile's Easter Island. Then, in May 2015, it entered Brazil, and all hell broke loose.

Who is at risk of contracting Zika? *Everyone* who hasn't previously been bitten by a Zika-infected *Aedes* mosquito. That means *billions* of people, most of whom live in the tropics.

Because of the risk of abnormalities in their unborn children, pregnant women are advised to avoid areas of the world where Zika has been reported. Women should also avoid *becoming* pregnant when in such areas. (However, because *Aedes* mosquitoes don't like the cold, it appears to be okay for pregnant women to travel to Mexico City and other places at high elevations, where temperatures drop into the forties or lower at night.)

By far the most common way to catch Zika virus is through the bite of an infected mosquito. However, Zika can also be sexually transmitted, including via oral sex. Before Zika, a mosquito-borne infection that can be transmitted sexually and cause birth defects was unprecedented. (The WHO recommends that males traveling to affected countries practice safe sex. And if they become infected and have a partner who is pregnant, they should abstain from sex for the duration of the pregnancy.) Other possible means of transmission are through breastfeeding by infected mothers and via transfusion with contaminated blood.

How will the Zika epidemic play out? No one knows. As of December 2016, the WHO reported mosquito-transmitted Zika virus infections in sixty-nine countries or territories and person-to-person-spread infections in thirteen of them. In North America, by mid-September 2016, 3,176 cases of Zika were reported by the Centers for Disease Control and Prevention (CDC). Most of these cases occurred in travelers to other countries, but forty-three infections were transmitted locally in Florida by *Aedes aegypti*. Of the 731 cases reported in pregnant women, a small number of infections were acquired sexually. And about one in ten U.S. Zika-infected pregnant women gave birth to a baby with birth defects.

In the United States, Zika virus infections first arrived in territories such as Puerto Rico, where by September 2016, 22,358 cases were reported, including 1,871 pregnant women. Pediatricians in Puerto Rico consider Zika to be a developmental doomsday virus.

In the continental United States, the first case transmitted from a locally infected mosquito was detected in Miami in July 2016. Research at that time suggested that fifty cities in the United States may be at risk for potential Zika virus outbreaks. The cities most at risk are in the southeast, especially Florida, as well as up the East Coast as far north as New York City. Public health experts feared a "perfect storm" when travelers from the Olympic games in Brazil would transport the virus to other areas of the southern United States. Some American Olympians did come home with other mosquito-borne viruses, but none were infected with Zika virus.

By early summer 2016, the Zika outbreak had peaked in Brazil. And in November, the WHO declared an end to the global health emergency status of Zika. However, the Zika crisis was by no means considered over. The WHO's message was Zika is here to stay, and so is their response to the crisis. (The situation is similar to what occurred with West Nile virus infection in the United States, when an epidemic that began in New York City in 1999 rapidly spread throughout the country and then became established as the main cause of viral encephalitis every year. You will read more about the West Nile virus pandemic in the next chapter.)

In the United States, public health officials predicted that the Gulf Coast of Florida would become the new ground zero for Zika. While some scientists predicted the Zika outbreak in Florida would be small and finished by winter, other experts suggested that it would be at least two years before the outbreak would wane in the Gulf states. In fact, by the end of 2017, only a small handful of Zika cases that had been acquired from locally infected mosquitoes were reported in Florida. And, as of October 2018, only fifty-two cases in Americans had been reported—and all were people who had traveled outside of the United States, where they almost certainly contracted the illness. This prompted some people to ask, What happened to Zika? Is this emerging infection now a submerged infection that might reemerge later? Only time will tell.

The areas of risk track with the presence of *Aedes* mosquitoes. While *Aedes aegypti* is found entirely in southern states, the more cold-tolerant *Aedes albopictus* has a foothold as far north as southern Minnesota. Based on where these species are found, residents of forty-one states are considered at risk.

As if the extraordinary spread of Zika virus in the Americas weren't enough to keep epidemiologists awake at night, in the summer of 2016 the virus invaded Southeast Asia. By September, 356 cases of locally acquired infection were reported in Singapore, and the CDC had issued a travel warning for pregnant women to eleven countries in the area, including India, Bangladesh, Thailand, and Vietnam. Before venturing to this area of the world, you should visit the CDC's travel website (wwwnc.cdc.gov/travel) for up-to-date information regarding the countries to avoid if you are pregnant or considering getting pregnant. Genetic studies suggest that the Southeast Asian and American strains of Zika virus are somewhat different.

The Enemy, Its Targets, and the Aftermath

Like dengue virus, Zika is a single-stranded RNA flavivirus. Even though Zika was first identified in a monkey in Uganda in 1947, it is unclear how (or when, or if) the virus jumped from primates to humans. Also, the role animals (other than mosquitoes) play in the transmission of the virus is unknown. Genetic analysis of Zika virus suggests that it was probably introduced to Brazil by an infected traveler from the South Pacific.

Just as with dengue virus, 80 percent of the people infected by Zika virus have no symptoms. Those who do become sick usually suffer for less than a week with fever, muscle and joint pains, conjunctivitis (red eyes), and a rash. No cases of bleeding have been reported. And until the spread of Zika to the Americas, no deaths were recorded. In early 2016, seven fatalities were reported in Venezuela, Colombia, and Brazil. In July, a seventy-three-year-old man in Utah became the first Zika casualty in the United States. He acquired his infection while traveling in Mexico. Of serious concern, it appears that he passed the virus on via his sweat or tears to his thirty-eight-year-old son caregiver.

If it weren't for the potential damage to fetuses and newborns—as well as a rare neurological complication known as Guillain-Barré syndrome [4] —Zika virus would be nothing more than a major nuisance. Instead, it is turning out to be an ever-spreading tragedy.

A number of studies have now characterized the types of brain abnormalities that occur in congenital Zika virus infection, some of which are unique to Zika. When it enters the brain, the virus appears to target microglia, cells that ordinarily defend the nervous system. This sets off a cascade of inflammatory mediators that damage other brain cell types, such as neurons. Results of several studies since 2015 suggest that from 5 to 7 percent of babies born to mothers with confirmed Zika infection had evidence of Zika-related birth defects.

Treatment and Prevention

Like dengue and chikungunya, there isn't a specific drug that can treat Zika. The current mainstay of therapy is the treatment of muscle and joint pain with nonsteroidal anti-inflammatory drugs. A vaccine isn't currently available, but an urgent rush is on to develop one; in January 2018, the U.S. Food and Drug Administration granted "fast track" status to a vaccine produced by a Japanese company. In the meantime, the same preventive measures outlined above for dengue and chikungunya are recommended.

And most sadly, for the serious complications of Zika virus infection—congenital brain disease and Guillain-Barré syndrome—there are also no current treatments.

As more is learned about the persistence of Zika virus in semen, the CDC updates its guidance regularly on how to protect yourself during sex. Their current recommendations are readily available online.

To prevent potential Zika infection by blood transfusion, the American Red Cross has asked people not to donate blood within twenty-eight days of returning from an affected area. And for countries such as Puerto Rico and the U.S. Virgin Islands, where Zika virus has already gained a solid foothold, the FDA has advised blood banks to import blood from Zika-free areas. In the fall of 2016, the FDA began recommending that all blood and blood products in the entire United States be tested for Zika virus.

Conventional methods of spraying with pesticides to eliminate mosquitoes have failed to stop the spread of all three arboviruses discussed in this chapter. More innovative approaches to targeting the *Aedes* mosquitoes are needed. One such approach, currently being tested, involves releasing sterile male mosquitoes—mosquitoes with reproductive systems damaged by radiation. They will mate with females but produce no offspring.

Perhaps most promising of all, researchers are testing a particularly potent type of genetic engineering called a "gene drive." Gene drives are sequences of DNA created in the laboratory that defy the rules of genetics by being able to copy themselves, so instead of half of the offspring inheriting the gene drive, almost all of them do. That is, they *drive* a desired genetic change, and after a short amount of time they can transform an entire wild population—in this case mosquitoes, like those that transmit Zika, dengue, and chikungunya, and even anopholine mosquitoes that carry the malaria parasite.

Another innovative approach is the release of genetically altered male mosquitoes that contain a gene that kills their offspring before they reach adulthood. A British company, Oxitech, has received tentative FDA approval for this technology. And in October 2016, the U.S. Agency for International Development awarded Delaware-based WeRobiotics a $30 million grant to develop drones that can carry sterile male mosquitoes into hard-to-reach areas. These sterile insects can greatly reduce populations of *Aedes aegypti*, thereby limiting the spread of Zika virus, as well as dengue virus and chikungunya virus.

Also, trials are underway in Australia in which *Wolbachia*-infected mosquitoes are released, using the same rationale for eradicating the *Aedes aegypti* mosquito mentioned earlier for combating dengue. And in 2017, the Florida Keys Mosquito Control District released twenty thousand male mosquitoes infected with *Wolbachia*. (In case you're wondering, the company MosquitoMate, mentioned earlier in this chapter, makes its money by raising and infecting these mosquitoes.)

As mentioned earlier, the race is on for the development of a Zika vaccine. According to the WHO, more than sixty research institutes and companies are now working on products to combat the spread of Zika virus. In August 2016, the NIH launched the first clinical trial of a Zika vaccine that looks highly promising. Nonetheless, a safe and effective vaccine is likely to take several years to develop. Some authorities predict one won't be available until 2020. Sadly, it may come too late for the outbreak sweeping Latin America and the Caribbean and now entering the United States and Asia.

Also, concern has arisen because of the experience with the dengue vaccine. It is possible that, as with the dengue vaccine, vaccination of young children with a Zika vaccine could make them susceptible to developing more severe disease when they become infected.

A final aspect of the mounting tragedy of Zika virus infection is its impact on unborn children and their families. Unfortunately, over 2,500 babies with microcephaly and other forms of brain damage will be born in Brazil alone because of the Zika epidemic. These babies are likely to suffer serious long-term learning disabilities requiring chronic care. Former WHO director-general Margaret Chan recently said publicly that planning is urgently needed to deal with the impact this tragedy will have on many nations' healthcare systems.

Lessons for the Future

"Lesson: We're only as safe as the most vulnerable people in the most vulnerable places in the world."—Médecins Sans Frontières

Still stinging from its highly criticized late response to Ebola, the WHO was quick to declare Zika a public health emergency. It was equally quick to call for more funding—some $25 million for its own efforts alone. Margaret Chan has championed the cause, but so far her calls for ramped-up financing of public health measures, medical care, and research have yielded inadequate results.

Leaders of the CDC and NIH have also pleaded for emergency funding, insisting that billions of dollars are required. In February 2016, president Obama asked the U.S. Congress for $1.8 billion to combat Zika. After much political wrangling, $1.1 billion was finally appropriated in October. The White House had to redirect funds for fighting the Ebola virus to support efforts to address Zika—but, of course, this is robbing Peter to pay Paul.

One of the most immediate research priorities is the development of a simple test to distinguish Zika, dengue, and chikungunya virus infections. Because of the overlap of the geography of these infections and the similarity of their symptoms, a test that rapidly sorts them out is urgently needed.

As already mentioned, a top priority is development of a Zika vaccine specifically for pregnant women. But this won't be easy, as there are several scientific and ethical barriers to developing vaccines for pregnant women.

Following on the heels of the Ebola epidemic, the Zika pandemic has highlighted the need for a rapid and coordinated effort by the WHO, CDC, NIH, and other organizations and enterprises. In March 2016, just such an idea was put forward by members of the Global Health Risk Framework Commission and the National Academy of Sciences, Engineering, and Medicine. An incremental $4.5 billion per year for spending on health systems, emergency response, and research was proposed. In the long run this investment would save millions of lives. Could Zika be the tipping-point pandemic that spearheads such a crucial development?

9

MICROBES IN FLIGHT

Birds and Bats

"If the Almighty were to rebuild the world and asked me for advice, I would have English Channels 'round every country. And the atmosphere would be such that anything which attempted to fly would be set on fire."—Winston Churchill

If Winston Churchill had his way, there would be no need for this chapter on microbes carried by birds and bats—or for any discussion on mosquito-borne illnesses.

For that matter, there would be little need for this book, because many of our microbial mortal enemies are transported around the world by planes carrying infected travelers. The ability to fly—whether you're a bird, a bat, an insect, or an airline passenger—has been central to the spread of microbes that can kill us or make us ill.

WEST NILE VIRUS

"Bad birds seldom bring good weather."—Icelandic proverb

The West Nile Pandemic

West Nile virus has been with us for at least eight decades. But it is considered an emerging infection because of the recent dramatic shift in its geographical range.

As its name suggests, West Nile virus was discovered in the West Nile region of Uganda, in 1937. Over the decades, it has caused epidemics in Africa, Israel, several European countries, and Russia. However, until the 1990s, West Nile disease was considered a minor problem for humans. Then West Nile–related outbreaks of a brain infection (encephalitis) were reported in Algeria and Romania. And when the virus arrived in New York City, this was a game changer.

West Nile virus was first identified there in the summer of 1999 by Tracey McNamara, chief pathologist at the Bronx Zoo. She was struck by the number of dead birds on the zoo grounds, including crows and flamingoes. At about the same time, doctors reported several fatal cases of encephalitis in patients in the neighboring borough of Queens. McNamara made the mental connection between the bird and human infections.

Soon afterward, the West Nile virus literally took off, infecting and killing a large number of birds—as well as humans and other animals—throughout the United States.

We don't know exactly how West Nile virus made its way to America. Most likely it got here via an infected migratory bird, or a stowaway mosquito on a flight from another country. But within a couple of years it became the most common mosquito-borne pathogen in the United States.

At first all the cases of West Nile virus were on the East Coast—but it quickly spread, via infected migratory birds, to all forty-eight states on the U.S. mainland. It is still very much with us.

The number of West Nile infections varies from year to year. In 2012, one of the worst years, 286 people died, with the state of Texas hit the hardest. By the end of 2015, 49,937 cases of West Nile disease, and 1,911 deaths from it, had been reported to the Centers for Disease Control and Prevention (CDC). But only 2,544 cases were reported to the CDC in 2018.

Like the three viruses you read about in chapter 8, West Nile virus is an arbovirus transmitted by infected mosquitoes—and then by other creatures that those mosquitos bite. The disease it causes shares some of the features of dengue, chikungunya, and Zika. But there's a marked difference between West Nile virus and other arboviruses: the West Nile microbe is far more promiscuous.

While dengue, chikungunya, and Zika are carried by only two species of mosquito—*Aedes aegypti* and *Aedes albopictus*—West Nile virus is harbored by at least sixty-five different mosquito species (including these and other *Aedes* species, as well as by *Anopholes* mosquitoes, the insects that carry the parasite that causes malaria).

But the main carriers of West Nile virus are mosquitoes belonging to the genus *Culex*. There are multiple *Culex* species. *Culex tarsalis* is the species most responsible for carrying West Nile virus to residents of Colorado; *Culex pipiens* is particularly nasty because it bites both humans and birds.

A second unique feature of West Nile virus is that its promiscuity extends to a broad range of animal victims, particularly birds. The virus has been isolated in more than 250 different bird species.

Some birds of the order Passeriformes, such as sparrows, are reservoirs for the virus. This means that when a sparrow is bitten by an infected mosquito, West Nile virus doesn't kill it. Instead, the virus grows steadily—and, for the sparrow, harmlessly—inside the bird. These microbes can then be readily transmitted via mosquitoes to other creatures, including human beings.

Other birds belonging to the Corvidae family, such as crows, ravens, and blue jays, aren't so lucky. These birds are dead-end hosts, which means that West Nile virus doesn't just grow inside them—it kills them. As a result, since 1999, some areas of the United States have witnessed a devastating loss of many common backyard birds, including the American robin, the house wren, the chickadee, and the tufted titmouse.

At least twenty-six mammal species, including humans and horses, are potential dead-end hosts for the virus. Even reptiles such as crocodiles aren't spared. (It's hard to imagine how mosquitos can penetrate this reptile's thick hide, but they manage to.)

The Enemy, Its Targets, and the Aftermath

Like dengue and Zika, West Nile virus belongs to the *Flavivirus* genus. There are more than seventy other flaviviruses; the most notorious are yellow fever, Japanese encephalitis virus, and St. Louis encephalitis virus, all of which can cause life-threating encephalitis. (When West Nile virus was first identified in a feverish Ugandan woman by Kenneth Smithburn in 1937, he found that it was related to yellow fever virus. And when West Nile virus first showed up in the United States, it was initially confused with St. Louis encephalitis virus.)

Like other mosquito-borne viruses, West Nile virus is transmitted only by the bite of an infected female mosquito. Fully 80 percent of people infected with West Nile virus have no symptoms at all. Of those who do show symptoms, over 99 percent have no neurological problems. These folks have what is called West Nile fever. West Nile fever's incubation period—the time between initial infection and the appearance of symptoms—is two to fifteen days. People with West Nile fever commonly complain of fever, headaches, fatigue, muscle pains, nausea, vomiting, and, sometimes, a rash. Typically, this illness lasts from five days to a month.

Like Zika, West Nile virus can attack nervous tissue—though it rarely does. Less than 1 percent of cases result in neurological disease—or, when the brain is involved, encephalitis. When this happens, the infected person often suffers severe headaches; they may also develop a stiff neck, muscle weakness, or mental confusion. This is the form of West Nile disease that is potentially fatal.

In rare cases, inflammation of structures in the nervous system can occur. When this involves the covering of the brain, the illness is called meningitis.

A very small number of infected people develop acute weakness—or even paralysis—in their arms or legs. Because these symptoms mimic those of polio, this form of the disease is called West Nile poliomyelitis.

Occasionally, people infected with West Nile virus may exhibit symptoms very similar to those of Parkinson's disease—tremors, muscle rigidity, dizziness.

People over the age of seventy have the greatest risk of a West Nile–related neurological illness, probably because the body's immune defenses wane with age (a process called immunosenescence). This idea is supported by the increased risk of West Nile virus neurological disease for people who have a compromised immune system, such as organ transplant recipients.[1]

Originally, the estimated fatality rate of 4 percent was attributed mainly to those with encephalitis who were over seventy years of age. But recent studies revealed that the fatality rate may be much higher because the West Nile virus infection leaves people who initially recovered more vulnerable to other infectious diseases and kidney problems. And researchers at Baylor University recently reported that nearly half of their patients who recovered eventually developed worsening neurological problems during the next decade.[2]

At first it was generally thought that being infected by West Nile virus—with or without symptoms—led to immunity. But a recent Israeli study suggests otherwise. In this study, fifty patients with evidence of a previous West Nile virus infection later developed a recurrent infection of the same disease. Furthermore, these patients had a much higher chance of developing a neurological illness—or of dying from the virus. Another disturbing finding was that, in some people, the initial virus later reactivated. The study also suggested that, in some cases, the virus could cause a psychiatric disorder.[3]

Treatment and Prevention

As with all the mosquito-borne viral infections, there is no antiviral therapy for West Nile virus. There have been some promising trials but nothing conclusive or worth recommending.

So far, the best treatments for people with the neurological form of West Nile illness are nonsteroidal anti-inflammatory agents. Still, a high percentage of infected people suffer for months or years from brain damage caused by the virus.

Similarly, a vaccine to prevent West Nile virus infection in humans doesn't yet exist. However, the trial of a potential vaccine, funded by the National Institute of Allergy and Infectious Diseases, began in 2015, and the early results are promising. A vaccine using killed West Nile viruses is already available for horses, and some zoos give this vaccine to their birds. But, as of this writing, its effectiveness is unknown.

The best way to protect yourself from West Nile virus is to avoid mosquitoes. Use mosquito spray with DEET, especially if you are over seventy years old or an organ transplant recipient; wear long-sleeve shirts and full-length pants when you're in areas where there are likely to be lots of mosquitoes; and get rid of all containers that can serve as breeding grounds for mosquitoes.

Because West Nile virus can be transmitted by blood transfusions, blood banks in the United States routinely screen for the virus.

Lessons for the Future

"Denial ain't just a river in Egypt."—Mark Twain

Unfortunately, in recent years the pharmaceutical industry has become less interested in developing a possible anti–West Nile virus drug. Research funding for combating West Nile virus has also declined.

Scientists recently found that the number of West Nile infections can be roughly predicted from year to year by tracking weather patterns. Not surprisingly, when temperatures in the previous year are above average, a bumper crop of mosquitoes follows. So, as the world warms, we will likely see more people with mosquito-borne illnesses of all types. Increased rainfall also favors the breeding of mosquitoes. (You will read more about the effects of climate change on vector-borne diseases in chapter 20.)

Weather-related events can also impact the populations and migratory patterns of birds. As we saw in chapter 5, everything is connected.

BIRD FLU

"The arrival of pandemic influenza would trigger a reaction that would change the world overnight. A vaccine would not be available for a number of months, and there would be very limited stockpiles of antiviral drugs. Foreign trade and travel would be reduced or even ended in an attempt to stop the virus from entering countries. It is likely that transportation would also be significantly curtailed domestically, as smaller communities sought to keep the disease contained."—Michael T. Osterholm

Bird Flu Outbreaks, Epidemics, and Pandemics

You're already familiar with influenza, commonly called the flu. You might also know that there is a connection between bird (avian) and human influenza that keeps public health experts awake at night. But the terminology can get confusing, so let's start with some clarifications.

Avian or bird flu is influenza caused by viruses that live in birds. The viruses of greatest risk—the ones that worry public health officials the most—are called highly pathogenic avian influenza (HPAI). These viruses can wipe out an entire flock of chickens within forty-eight hours. One such HPAI strain, H5N8, hit the Midwestern United States in 2014–2015. Close to fifty million chickens and other commercially raised birds were killed to control the epidemic. And the economic loss in the United States alone was estimated to exceed $3 billion.

Fortunately, no human transmissions of H5N8 were reported. But what most worries public health professionals is the ability of HPAI strains to cross over to people. When this happens and the virus becomes capable of person-to-person transmission, pandemics of severe influenza can be spawned. As Michael Osterholm, a leading infectious diseases epidemiologist, warns convincingly (and alarmingly) in his book *Deadliest Enemy: Our War against Killer Germs*, "As infectious disease epidemiologists, we all know that pandemic influenza is the one infectious disease that *will happen*."

Fortunately, most influenza viruses that are adapted to birds don't infect humans. Although they circulate widely in birds, especially waterfowl, they don't usually even cause any symptoms. But the ones that do cause illness can be devastating to bird populations. Outbreaks of avian influenza in chicken and turkey populations are a constant threat to the poultry industry.

But some strains of avian flu can jump to other species, including *Homo sapiens*. The most notorious spillover of an avian virus to humans occurred in 1918–1919. This resulted in a flu pandemic in which 20 to 40 percent of the world's population became ill, and an estimated fifty to one hundred million people died. In 2018, we remembered and reflected on the centenary year of the 1918 pandemic—the deadliest event in U.S. history.

Influenza viruses are divided into three types: A, B, and C. Types A and B are the ones that cause annual or seasonal flu epidemics. They typically sicken up to 20 percent of the population in any given year. Influenza A viruses are found mainly in wild birds, but also in humans, pigs, horses, and even whales. Influenza B viruses also circulate widely, but only among humans.

Here's where it gets potentially confusing. Influenza A viruses are divided into multiple subtypes, based on two proteins on their surface: hemagglutinin (H) and neuraminidase (N). There are sixteen different hemagglutinin subtypes and nine different neuraminidase subtypes. All known subtypes of influenza A viruses have been found among birds, except for two that are only found in bats. The problem is that influenza A viruses can suddenly mutate—and cause big problems.

As Margaret Chan, former director-general of the World Health Organization, observed, "The unique nature about influenza virus is its great potential for change, for mutation." When a new influenza A virus emerges, a flu pandemic—a global disease outbreak—can occur. Unlike epidemics of seasonal flu, which occur every year, flu pandemics are relatively uncommon—but much more potentially deadly.

Influenza epidemics have been around for centuries. A description of flu symptoms can be found in the writings of Hippocrates, 2,400 years ago. The first epidemics that we're almost certain were influenza were reported in Europe in the sixteenth century. Severe epidemics also occurred in the eighteenth and nineteenth centuries.

In the twentieth century we had three flu pandemics. Of these, the 1918–1919 Spanish Flu—referred to by influenza researchers Jeffrey Tautenberger and David Morens as "The Mother of All Pandemics"— was the most devastating.

It is likely that this pandemic began in the United States. But because Spain was neutral in World War I, news of the disease decimating American troops wasn't suppressed in that country, as it was in American media. This led to the mistaken label "Spanish Flu." The pandemic went on to kill more people than all the wars of the twentieth century combined.

Avian flu virus subtype H1N1 was the culprit. In 1957–1958, H2N2 emerged and was responsible for the Asian Flu pandemic. And in 1968–1969, H3N2 triggered the Hong Kong Flu pandemic.

So far in the twenty-first century, only one flu pandemic has materialized—in 2009–2010. This pandemic was caused by a variant of the H1N1 virus (dubbed H1N1v), which jumped from pigs to humans, and thus became known as swine flu. (Later, contact with pigs at state fairs or live animal markets was recognized as a source of other subtypes of swine flu.) Because pigs can simultaneously harbor flu viruses from both birds and humans, as well as their own flu strains, they can serve as melting pots for genes from many different influenza viruses—a phenomenon called reassortment. This is one of the ways new flu viruses are created.

In recent years, several flu viruses have been on public health officials' radar screens. Of these, subtype H5N1—an HPAI virus—emerged in a Hong Kong poultry market in 1997. Since then, this virus has spread from Asia to Europe and Africa, resulting in many millions of infected and dead birds. This has done great harm to the livelihoods of hundreds of thousands of people—and to the economies of a dozen countries. By early 2016, 850 human cases of H5N1 infection had been reported. Nearly 450 of the infected people died—a shockingly high mortality rate of 53 percent. (In comparison, the mortality rate of the H1N1 virus in the 1918–1919 pandemic was only 2.5 percent.) Fortunately, the illness has not yet reached the Americas.

In 2013, flu subtype H7N9 was found in poultry in China. As of this writing, about 1,600 H7N9 infections had been reported in humans, and 40 percent of these people died from their infections. Because the extraordinary surge of H7N9 infections had overtaken the number of human cases of H5N1 in a short time period, public health officials are understandably on edge.

To make matters even worse, the first human case of a new bird flu virus, H7N4, was reported on Christmas Day in 2017. The patient was a sixty-eight-year-old woman hospitalized in southern China. She had picked up the virus from a chicken. As I write this paragraph in June 2019, it is too early to know what this bird flu virus has in store for us.

Notably, in virtually all human cases of H5N1 and H7N9 infections, the viruses were picked up through close contact with infected poultry. Fortunately, neither virus has yet acquired the ability to spread from human to human, as occurred in the three twentieth-century pandemics.

While concern about the emergence of new bird flu pandemics captures people's attention, let's not forget that ordinary, seasonal flu remains a big health problem. For example, each year, 5 to 20 percent of all Americans get the flu, and more than two hundred thousand are hospitalized because of seasonal flu-related complications. Between 1976 and 2006, annual deaths associated with seasonal flu ranged from a low of three thousand to a high of forty-nine thousand. In 2018, seasonal influenza was the worst in over a decade, killing eighty thousand people in the United States. Many hospitals were filled to capacity, and some resorted to using "surge tents" to handle the overflow of patients. The number of fatal cases of flu in the 2018–2019 season was fifty-seven thousand as of April 2019—still quite high.

Here's something else worth noting: seasonal flu takes its biggest toll on children under five, the elderly, and people with underlying medical conditions—but pandemic flus tend to sicken young, healthy people. The

1918–1919 flu pandemic killed *mostly* young people who had been healthy. This pandemic was so large, and so swift, that life expectancy in 1918 dropped by twelve years. Nobody knows for sure why young people were (or are) at increased risk of dying from avian flu pandemics.

The Enemy, Its Targets, and the Aftermath

Given the long history of flu pandemics—and the similarity of bird, swine, and human flus—scientists think that human influenza probably originated when humans first began domesticating animals. But the discovery of a biological link between flu in animals and flu in humans did not occur until 1918, when veterinarian J. S. Koen observed a disease in pigs that he proposed was the same as the infamous 1918–1919 flu pandemic in humans.

In 1918, most physicians and scientists mistakenly believed that influenza was caused by a bacterium called Pfeiffer's bacillus. Science hadn't yet recognized that this bacterium actually caused life-threatening pneumonia, a common complication of influenza. (This bacterium has since been renamed *Haemophilus influenzae*.)

In 1928, Robert Shope at the Rockefeller Institute for Comparative Pathology inoculated healthy pigs with filtered fluid taken from pigs that had swine flu and was able to reproduce the disease. This provided the first reliable evidence that influenza was caused by a virus. (As a result, however, H1N1 flu was initially—and mistakenly—believed to have come from pigs, not birds.)

Influenza viruses are members of the Orthomyxoviridae family. Like HIV, influenza virus has a paltry number of genes—only eight. (You'll recall that we humans have over twenty-one thousand genes.) But influenza virus genes are constantly changing their makeup, giving rise to mutant viruses that can evade both the body's immune system and vaccines.

For a flu virus to thrive in animals, it must have the ability to reproduce in the cells of the animal's respiratory tract—its sinuses, throat, or lungs. Transmission to other animals, including humans, occurs either through direct contact or when droplets of fluid containing the virus move from one creature to another (for example, via a sneeze or cough).

The complex interplay of factors that drive pandemic flus are not yet well understood. For example, it isn't clear why the H1N1, H5N1, and

H7N9 subtypes are so virulent. But it appears that these viruses provoke a massive release of proteins called cytokines in our immune system. (This phenomenon, called a cytokine storm, in turn triggers a release of other inflammatory molecules, which then damage vital organs such as the lungs and kidneys.)

It also isn't clear why H1N1 was so easily transmitted from person to person, while H5N1 can't spread between humans at all. Furthermore, nobody knows if a mutation of H5N1 or H7N9 will (or even could) develop that would allow it to be spread among people. If this *were* to happen, however, it would be catastrophic, given the high mortality of H5N1 and H7N9.

All flus are infections of the respiratory tract. The most common symptoms are a sore throat and a dry cough, accompanied by fever, headache, muscle pains, and fatigue. Sometimes gastrointestinal symptoms, such as nausea, vomiting, and diarrhea, occur as well. (So-called stomach flu is usually not a flu at all but is caused by some other type of virus.)

When an influenza virus enters the lungs, it often either causes viral pneumonia or allows a secondary bacterial pneumonia infection to take hold. When one of these sets in, people can become critically ill and require hospitalization.

Distinguishing the flu from a common cold, which is caused by other types of respiratory viruses, can sometimes be difficult. A flu is usually more—sometimes much more—severe. Flu is also more likely to develop during the colder half of the year, which is why those months are called the flu season in the northern hemisphere.

Treatment and Prevention

Here is the single most important thing you need to know about influenza: if you think you might have it, contact your doctor promptly. They can help you decide what treatment is best.

Antiviral drugs can successfully treat the flu—but they must be started promptly after symptoms first appear. Children, the elderly, people who are very sick, and those who have underlying medical conditions should be given antiviral drugs without waiting for confirmation of the virus through laboratory tests.

As of 2019, four FDA-approved antiviral drugs are recommended by the CDC for treating the flu. The CDC's website on treatment of influenza is an excellent resource. (Because new drugs appear all the time and the CDC regularly updates its recommendations, I won't list the drugs here. But one new agent, baloxavir, is particularly interesting because it is given in a single dose.)

Even when antiviral drugs are started promptly, however, they don't always work. Some flu viruses are drug resistant. But the pharmaceutical industry is at work on new agents—and there is certainly a big market for them.

Prevention of pandemic avian flu in humans usually starts with the slaughtering of any bird flocks that appear to harbor it. For example, in 2008, many millions of birds were killed on poultry farms in or near Hong Kong, in two waves. The first occurred after a routine check of fecal samples uncovered H5N1. The second took place after the virus killed scores of chickens at a chicken farm.

Although personal hygiene (including hand washing and covering your mouth and nose when you cough or sneeze) has a small but tangible protective effect, vaccination is the mainstay of flu prevention. This vaccination covers many different types of flu, and its makeup changes from year to year. Each year, before flu season begins, experts determine which flu viruses to address in the annual flu vaccine. Sometimes their predictions are good. For example, the 2015–2016 influenza vaccine worked well and was one of the reasons why 2015–2016 was a mild flu season. On the other hand, the vaccine used for the 2014–2015 flu season was ineffective, probably due to mutations of the viruses that it targeted. Consequently, 2014–2015 was a severe flu season, especially for people over sixty-five years old. Similarly, because the H3N2 virus mutated after it was included in the flu vaccine, the vaccine that was rolled out in 2017 was only about 25 percent effective.

Even under the best of circumstances, however, flu vaccines only protect us about 60 percent of the time. While the CDC reported in 2017 that it had created a candidate vaccine for H7N9 avian flu, there remains a great need for more effective vaccines. (More on this topic in chapter 19.)

Some encouraging news came from the U.S. Congress in 2018, when a legislative proposal surfaced that would fund development of a universal influenza vaccine—that is, one that would protect us from *all* flu strains, including avian flu. In April 2018, Bill Gates announced a donation of $12 million to help create a universal vaccine. If this goal were to be realized, it would be an extraordinary—and an extraordinarily life-saving—achievement.

Lessons for the Future

"For a pandemic of moderate severity, this is one of our greatest challenges: helping people to understand when they do not need to worry, and when they do need to seek urgent care."—Margaret Chan

By their very nature, bird flu pandemics are international problems that require coordinated international responses by both governments and organizations such as the WHO, the CDC, and the National Institutes of Health (NIH). In addition, the pharmaceutical industry, the main supplier of vaccines and antiviral drugs, needs to be at the table.

You'll recall from my discussion of Zika in chapter 8 that the creation of such a coordinating body was recently proposed by leaders of the Global Health Risk Framework Commission and the National Academy of Sciences, Engineering, and Medicine. Should this body be set up, preparing for flu pandemics would almost certainly be on its front burner.

Of the many challenges posed by pandemic flu, the most disturbing one is our inability to predict when a pandemic will occur. As Hugh Pennington, emeritus professor of bacteriology at the University of Aberdeen, has observed, "Predicting influenza pandemics and their impact is a fool's game." Nonetheless, we need highly qualified planners to do their best at preventing and managing both seasonal and pandemic flus.

As pointed out by the distinguished influenza researchers Wenqing Zhang and Robert Webster in the journal *Science* in 2017,[4] the fact that we lack the fundamental knowledge to predict if and when an influenza subtype will acquire pandemic ability was, and remains, particularly sobering as we then approached the one-hundred-year anniversary of the 1918 Spanish influenza—one of the greatest public health crises in history.

NIPAH VIRUS

> "Bats have no bankers and they do not drink and cannot be arrested and pay no tax and, in general, bats have it made."—John Berryman, American poet

Before we get to the dark side of bats, let's consider two facts about these marvelous creatures:

1. First, there are 1,240 known bat species worldwide. That's about 20 percent of all 5,416 mammal species.
2. Second, insect-eating bats contribute an estimated $3.7 billion to the North American agricultural economy every year. (These are the echolocating bats you see at dusk flitting about your backyard. A single bat can eat up to 1,200 mosquito-sized insects every hour—six to eight thousand each night—thus making your backyard more comfortable.)

Now the bad news. From the smallish insect-eating bats to the larger, fruit-eating bats called flying foxes, bats harbor sixty-six different species of viruses. Yet bats themselves rarely get sick from any of these viruses. (Zoologists think their remarkable resistance to viral disease is due to their fifteenfold increase in metabolism when flying, which raises their body temperature to levels that viruses can't tolerate.) Eight of these sixty-six viruses, including Nipah virus, can cause devastating infections in humans.[5]

Nipah Virus Outbreaks

Nipah virus was first recognized in Malaysia and Singapore in 1999, during an outbreak of encephalitis and respiratory illness among pig farmers and other people who had close contact with pigs. Its name originated from Sungai Nipah, a village where pig farmers were dying of encephalitis. It is related to Hendra virus, which causes encephalitis in horses and humans and is harbored in bats. Bats of the genus *Pteropus*, known as flying foxes, were quickly identified as the reservoir for Nipah virus.

While Nipah virus doesn't trouble flying foxes and causes only a mild illness in pigs, it can be deadly to humans. In the initial outbreak, nearly three hundred cases and over one hundred deaths were reported. To contain the outbreak, over a million pigs were slaughtered, causing tremendous economic loss for Malaysia. Fortunately, because humans' exposure to infected pigs was the main route of the virus's transmission, this strategy seems to have worked. (But Nipah virus can also be transmitted by direct contact with infected bats, where the virus lives on.)

In 2001, Nipah virus showed up again, this time in Bangladesh and India. Genetic testing revealed that the Nipah virus strain in these countries was different from the one in Malaysia. So far, close to three hundred cases of Nipah virus infection have been reported in India and Bangladesh. The fatality rate—40 to 70 percent—is similar to what occurred in Malaysia. However, with this strain there have been reports of person-to-person transmission of the virus in hospitals. Also, unlike the Malaysian outbreak, which occurred only once, outbreaks occur almost annually in Bangladesh.

Scientists have figured out how Nipah gets transmitted in that part of the world, and it doesn't involve pigs. Instead, people eat raw date palm sap that has been contaminated by flying foxes.

The Enemy, Its Targets, and the Aftermath

Nipah virus belongs to the *Henipavirus* genus. It attacks the brains of pigs and humans. Studies of the evolution of Nipah virus indicate that it first evolved in 1947, though we're not sure just where.

After an incubation period of five to fourteen days, the illness leads to fever and headache, followed by drowsiness, disorientation, and mental confusion. In the early stage of infection, some patients also have respiratory symptoms. Over half of the patients admitted to hospitals during the Malaysian outbreak had distinctive signs of illness in their brain stems. This is the part of the brain that regulates vital bodily functions. About one-third of these patients soon died; 15 percent suffered long-term neurological problems, such as seizures; and 53 percent recovered fully.

As of this writing, there are no antiviral drugs to treat Nipah virus, so treatment is limited to supportive care. Because Nipah virus can be transmitted from person to person, measures are needed to prevent infected, hospitalized people from spreading it to others. It also goes without saying that people in affected areas should avoid sick pigs and bats, and should not drink raw date palm sap.

A vaccine against the closely related Hendra virus became available for horses in 2012. Because it elicits antibodies that also protect against Nipah virus, it might soon be modified into a Nipah virus vaccine.

Lessons for the Future

> "People and gorillas, horses and duikers and pigs, monkeys and chimps and bats and viruses: We're all in this together."—David Quammen, American science and nature writer

The emergence of Nipah virus two decades ago is a frightening reminder that new infections will continue to show up out of nowhere. And a recent article in *Nature* by Kevin Olival and colleagues, "Host and Viral Traits Predict Zoonotic Spillover from Mammals," shines the spotlight on bats as the mammalian reservoir for viruses that can jump to humans.

But Nipah virus hasn't had nearly the impact on human beings as two other, older viral pathogens that also circulate between animals and people in Asia: rabies virus and Japanese encephalitis virus (JEV). Because these viruses have been around a long time and haven't changed their geographic reach, they're not considered emerging pathogens.

Rabies—which, like Nipah virus, is carried by bats—continues to kill an estimated fifty-nine thousand people per year. Rabies is spread by the bite of infected animals and is almost always fatal.

JEV, a mosquito-borne flavivirus similar to West Nile virus, is responsible for an estimated ten to fifteen thousand deaths a year. JEV isn't found in bats—but, as with Nipah virus, pigs are often the source of infection in humans.

The good news in our battles against both rabies virus and JEV is that highly effective vaccines are available to prevent them. But, just as with Nipah virus, there is an urgent need for *treatments* for these viruses. These viruses also highlight the enormous importance of interdisciplinary research teams that include veterinarians, ecologists, epidemiologists, and physicians.

SEVERE ACUTE RESPIRATORY SYNDROME (SARS)

"This syndrome, SARS, is now a worldwide health threat. . . . The world needs to work together to find its cause, cure the sick and stop its spread."—Gro Harlem Brundtland, director-general, World Health Organization (1998–2003)

"We have seen SARS stopped dead in its tracks."—Gro Harlem Brundtland

The SARS Pandemic

Very few pandemics as terrifying as SARS have emerged—or been contained—with such lightning speed. Between November 2002 and July 2003, an outbreak of SARS that began in China quickly spread to thirty-seven countries. It sickened 8,096 people and killed 774 of them (a 9.6 percent fatality rate). On March 15, 2003, the then director-general of the WHO, Gro Harlem Brundtland, and her team gave the newly recognized disease its name—SARS. They also issued a call for health authorities worldwide to work together to stop the disease.

On July 9, 2003, barely four months later, that mission had been accomplished, and the WHO declared the pandemic contained. Almost miraculously, no cases of SARS have been reported anywhere since 2004.

The unprecedented speed of both the spread and the containment of SARS was the result of modern technology. The first reported case of SARS was a farmer in Guangdong, China, in November 2002. After a much-criticized initial delay in reporting cases in the region, the Chinese government began to recognize the potential destructiveness of the outbreak.

Then, in February, an American businessman developed pneumonia-like symptoms while on a flight from China, and died in the French Hospital in Hanoi.

Reports of serious flulike symptoms reached the world spotlight in early 2003, disseminated via the internet by the WHO, the CDC, and other health agencies. In late March 2003, the cause of the pandemic was identified—a new coronavirus that was named SARS-CoV. This scientific and technological feat was accomplished by a coordinated effort involving teams of researchers in Hong Kong, the United States, and Germany.

The puzzle regarding the origin of SARS-CoV was also solved surprisingly quickly. In May 2003, live, captured wild animals for sale in local food markets in Guangdong were found to harbor the virus. In particular, palm civets—which look like a cross between a raccoon and a possum—were found to be infected. In a swift response, ten thousand of these otherwise innocent creatures were killed.

However, the search for the original animal reservoir for human SARS-CoV took somewhat longer. Early on, bats were suspected, and in 2013 the Chinese horseshoe bat was identified as the source of human SARS-CoV.

The way SARS-CoV was spread—through close person-to-person contact—and the method of its transmission—by respiratory droplets produced by coughing or sneezing—also became apparent early in the epidemic. In February, an outbreak began in the Metropole Hotel in Hong Kong. The initial case was an infected doctor from Guangdong province in China, where retrospective analysis indicated the pandemic originated. He was considered a "superspreader" because, during his stay in the hotel, he infected sixteen other hotel visitors, who then brought the virus with them when they boarded planes for Canada, Singapore, Taiwan, and Vietnam.

One of the most alarming aspects of SARS-CoV was its relative ease of transmission to hospital caregivers. With most illnesses, the risk of hospital workers picking them up from patients is low. But SARS-CoV was very much an exception. In one Toronto hospital, for example, 37 percent of the 128 people that it treated for SARS-CoV were their own hospital staff.

In 2004, I happened to visit a hospital affiliated with Hong Kong University, where a good deal of elegant work was being done on SARS-CoV. I was greatly impressed with the high level of infection control measures that had been implemented at the hospital, where some years earlier multiple staff members had acquired SARS, and some had died.

The Enemy, Its Targets, and the Aftermath

SARS-CoV is a single-stranded RNA virus belonging to the Coronaviridae family. Before SARS, coronaviruses were only associated with mild respiratory tract infections in humans. Along with the rhinoviruses, they are a frequent cause of the common cold. (Coronaviruses infect many other animals as well. In dogs, they cause a contagious upper respiratory tract infection called kennel cough.)

The initial symptoms of SARS resemble the flu—fever, sore throat, cough, and muscle pain. Some people also experience shortness of breath, which occurs when the infection has moved from the upper respiratory tract—the sinuses and throat—into the lungs. SARS-related pneumonia (involvement of the lungs) is the main cause of death.

Like severe influenza, SARS creates a cytokine storm, which is produced by an overly activated immune system. This can lead to severe organ damage or even death. Also like the flu, a secondary bacterial infection may take hold in the lungs. This, too, can be fatal.

Treatment and Prevention

As with all viral infections, antibiotics are ineffective in the treatment of SARS. However, if a secondary bacterial infection of the lungs develops, there are antibiotics that can help.

Because of SARS's combination of deadliness and infectiousness, anyone suspected of having the illness must be isolated—if possible, in a negative-pressure room. This is a room, found in some hospitals, that is designed to prevent air from flowing out of it. During the early months of the SARS pandemic, this and many other measures of infection control were ramped up at hospitals around the world.

At this time, no vaccine exists for SARS that is safe for humans. Fortunately, there is no need for a vaccine—for now.

Lessons for the Future

While the SARS pandemic was quelled amazingly quickly, the threat of SARS is likely still out there. In the initial 2003 investigation, the virus was found in manguts (small mammals also known as raccoon dogs or tanukis), ferrets, and domestic cats—and, of course, in palm civets. Some of these creatures may still harbor the virus. And the main animal reservoir of human SARS, the Chinese horseshoe bat, is still flying merrily around Asian skies. A recent study of thirty-five bat species in China revealed that about 6 percent of them carried one or more of ten different species of SARS-like coronaviruses. In November 2017, a group of Chinese virologists reported in the journal *PLOS Pathogens* that they had found *all* the genetic building blocks of SARS virus in a single group of horseshoe bats living in a remote cave in Yunnan province.

The SARS pandemic demonstrated that a swift and aggressive response to emerging infections is vital not only to world health but to local and national economies. Following the arrival in Toronto of an infected traveler from Hong Kong on February 23, 2003, the city was essentially shut down for a short time. The eventual cost to the Toronto economy: $1 billion. And the total cost of the SARS pandemic to the global economy is estimated at $54 billion.

THE WUHAN CORONAVIRUS

In January 2020, as I reviewed the page proofs for this book, Chinese authorities reported an outbreak of pneumonia in fifty-nine patients in the city of Wuhan to the WHO. The outbreak began in early December and was traced to an animal market that sold bats, marmots, snakes, and poultry. On January 11, the first fatality was reported. And, amazingly, that same day Chinese researchers reported that they had sequenced the genome of the virus—a new coronavirus named 2019-nCoV.

By January 22, a total of 555 cases had been confirmed, and the virus had spread to several cities in China as well as to Thailand, Japan, and South Korea. The first case in the United States was reported on January 21—a traveler arriving in Seattle from Wuhan.

On January 24, three Chinese cities (and 18 million people) were on lockdown, and surveillance had been implemented at five airports in the United States. Although the WHO had yet to declare a "public health emergency of international concern," the CDC had activated its emergency response system. The Chinese government began building a new hospital in Wuhan to better handle the outbreak—with plans to complete construction within *six days*. Although the world's public health experts came to grips with the SARS pandemic with amazing speed, the initial response to the Wuhan Coronavirus appears to have been even swifter.

10

DON'T BREATHE THIS AIR

"Water, air, and cleanness are the chief articles in my pharmacy."—
Napoleon Bonaparte

Everyone knows that air is a mixture of gases. But the air we breathe is also filled with airborne particles called bioaerosols. Bioaerosols are suspensions of airborne particles that contain microbes (primarily viruses) and solids and liquids they produce. You've already read about some very dangerous bioaerosols, such as smallpox, influenza viruses, and SARS-CoV.

Many of the viruses, bacteria, and fungi that cause human disease are transmitted from person to person through the air. They are usually expelled through coughs or sneezes. A single sneeze sends millions of tiny droplets into the air at a speed of about two hundred miles per hour. The expelled microbes may then travel alone as free particles or inside droplets for a length of most rooms.

A single droplet can contain tens of thousands of microbes. Remember, viruses are very tiny—about 0.02 to 0.30 micrometers in diameter. (There are over twenty-five thousand micrometers in an inch.) Bacteria are bigger but still usually no bigger than two micrometers in diameter.

This chapter highlights two emerging airborne pathogens—a new viral threat (MERS-CoV), and a bacterium that was identified about forty years ago in Philadelphia: *Legionella pneumophila*, the cause of Legionnaires' disease.

MIDEAST RESPIRATORY SYNDROME (MERS)

"Trust in Allah, but tie up your camel."—Arabian proverb

The MERS Epidemic

No sooner had SARS disappeared—and buoyed hopes that we had seen the last of deadly coronaviruses—than a patient died on June 12, 2012, in a hospital in Jeddah, Saudi Arabia. This was the first recognized case of another severe respiratory disease that became known as MERS. The cause was yet another novel coronavirus, named MERS-CoV.[1] The MERS epidemic underscores the enormous importance of geography in the genesis and evolution of epidemics. Although the first case was identified in Saudi Arabia in June 2012, an outbreak of thirteen patients was recognized retrospectively in Jordan in April of that year. Since then, cases have been identified across all countries in the Middle East. Of the MERS cases that have been diagnosed in seventeen countries outside the Middle East, all occurred in travelers returning from that part of the world.

By August 2018, 2,229 cases of MERS had been reported in twenty-seven countries, with a fatality rate of 35 percent. Saudi Arabia, where the disease was first recognized, accounted for 83 percent of the cases. Patients who are elderly or who have chronic lung disease, diabetes, kidney failure, or other serious health problems have an increased risk of fatal infection.

As with SARS, catching MERS typically requires close contact with an infected person. Also as with SARS, there have been multiple hospital-based outbreaks. The largest one so far outside of the Middle East involved 186 cases in sixteen hospitals in South Korea in 2015. This outbreak was characterized by five super-spreading events in hospitals. In one of these outbreaks, eighty-two individuals became infected after contact with a single patient in an overcrowded emergency room. Thankfully, that outbreak is now over.

Because the spread of MERS is uncommon outside hospitals, its risk to the global population is considered low. That's fortunate, given that the Hajj—the Muslim pilgrimage to Mecca—attracts two to three million people every year to a very small area in Saudi Arabia. So far, there have been no Hajj-related outbreaks.

While MERS shares many features of SARS—most notably the development of severe, life-threatening pneumonia—one striking difference involves the animals that harbor the viruses. Bats are the main culprit in supporting SARS-CoV, but MERS-CoV lives, thrives, and reproduces in dromedary camels, without doing them much or any harm. Researchers in Saudi Arabia, the country hardest hit by the epidemic, believe that MERS-CoV lived in dromedaries long before the virus mutated and jumped to humans. They also think that MERS-CoV has jumped from animals to humans more than once.

Exactly how the virus moves from camels to humans is unclear. In fact, relatively few MERS patients have a history of exposure to camels. There may be a link with the consumption of unpasteurized camel milk, which is a popular beverage in Saudi Arabia. How the disease moves from person to person is much clearer: MERS is transmitted primarily by airborne droplets. Contact with surfaces contaminated with MERS-CoV may also play a role.

The Enemy, Its Targets, and the Aftermath

MERS-CoV was discovered by Ali Mohamed Zaki, a prominent virologist working at a hospital in Jeddah, where the first case of MERS was identified. Like SARS-CoV, MERS-CoV is a single-stranded RNA virus belonging to the genus *Betacoronavirus*. It too has a strong affinity for the cells lining the airways of the lungs. A recent report from a team of Chinese researchers showed that MERS-CoV also infects and kills a type of cell in the immune system called T lymphocytes. (You may remember from chapter 4 that these cells play a pivotal role in adaptive immunity. This could be one reason why MERS-CoV is so dangerous and often deadly.)

After MERS incubates for up to twelve days, the illness begins. Some people have no symptoms at all; others have only mild respiratory problems, similar to the common cold. But other people get quite ill. Typically, infected people experience a fever, a cough, and shortness of breath. These can progress to pneumonia within a week. Gastrointestinal symptoms such as diarrhea can also occur. About half the people with the illness experience severe respiratory disease. As mentioned, over a third die. When MERS is fatal, the symptoms commonly include cardiovascular collapse and kidney failure, as well as lung disease.

Treatment and Prevention

Neither a specific drug treatment nor a vaccine currently exists for MERS. (In 2018, a clinical trial of the first vaccine, dubbed INO-4700 or GLS-5300, was reported to show promise.) Supportive care is the mainstay of therapy. When lung symptoms are severe, mechanical ventilation and support in an intensive care unit are necessary.

The main prevention-and-control measure is avoiding exposure to droplets by wearing a surgical mask. Because MERS, SARS, and other serious respiratory viral infections are primarily spread in hospitals, much attention is paid to the exact type of mask that hospital workers need to wear. (If you work in a hospital or are interested in the ins and outs of surgical mask selection, visit the web pages on MERS at the WHO, the CDC, or the Saudi Ministry of Health.) Wearing a gown and gloves when entering a MERS patient's room, and removing them on leaving, are also recommended. Patients with MERS should be isolated in negative-pressure rooms to help contain the virus.

Of course, don't drink unpasteurized camel milk. If you have an existing health condition that's serious or chronic, it's also wise to avoid camels in general.

Camels infected with MERS-CoV may develop rhinitis—a runny nose—or might show no signs of infection at all. The WHO, the CDC, and the Saudi Ministry of Health have developed safety guidelines for people who work with camels. (In the United States, camel meat has been appearing on more and more menus, and the commercial camel-raising industry, though still quite small, has been growing.)

Lessons for the Future

> "Everything has to do with geography."—Judy Martz, former governor of Montana

While much has been learned in the first few years of the MERS epidemic, many questions remain unanswered. Where exactly did MERS-CoV come from? What are the underlying mechanisms of immunity? Can an effective vaccine be developed? What about an antiviral drug? And, perhaps most important of all, will a mutation in the genes of MERS-CoV occur that will allow the virus to be readily transmitted from person to person through the air? If and when that occurs, we could have a serious pandemic on our hands.

After reading about viruses such as MERS-CoV, as well as the viruses discussed in chapter 8, you probably won't be surprised if, when you next check in at a clinic, hospital, or emergency room, you're asked, "Have you traveled anywhere outside the country in the past three months?" It's an absolutely essential question to ask. Indeed, in the early years of my career, my colleagues and I were often stunned by healthcare workers' lack of understanding of the role of geography in the spread of certain infectious diseases. Fortunately, today, every decent healthcare professional understands that a life-threatening infection can be only a plane ride away.

LEGIONNAIRES' DISEASE

"In the case of Legionnaires', persistent pressure from the news media, a number of health officials said later, helped hold them accountable and to spur scientists to do what they rarely had done in other unsolved cases and outbreaks—take a crucial second look that solved the Legionnaires' outbreak."—Lawrence Altman, *New York Times* science writer

The Legionnaires' Disease Epidemic

Like MERS, Legionnaires' disease is a lung infection—a type of pneumonia. And, like the virus that causes MERS, the bacterium that causes Legionnaires' disease, *L. pneumophila*, gains entry to the lungs as a bioaerosol. But in the case of Legionnaires' disease, the offending bacterium isn't spread through person-to-person transmission. Instead, the source of *L. pneumophila* is contaminated water.

Pneumonia is the eighth most common cause of death in the United States—and, for children, the single most deadly infectious disease in the world. While bacteria are the most common microbes that cause pneumonia, the vast majority of bacterial pneumonias, including Legionnaires' disease, aren't contagious. You can't catch them from another human being. (One notable exception is the bacterium that causes tuberculosis,

Mycobacterium tuberculosis, which is transmitted entirely from the coughs and sneezes of folks who have the illness.)

If you were alive in July 1976, which marked the two hundredth anniversary of the declaration of independence from Great Britain, you may recall the many festive celebrations throughout the country. But things were not so festive if you were among the four thousand World War II Legionnaires, and their families and friends, who assembled in Philadelphia for the fifty-eighth American Legion convention.

Shortly after the July 4 celebration, an outbreak of pneumonia occurred that was linked to the Bellevue-Stratford Hotel, where many of the conventioneers stayed. (This association led to the name *Legionnaires' disease*.) One hundred and eighty-two people came down with the illness, and twenty-nine died. Some of the infected people never actually entered the hotel, which means that the bacterium had spread into and through the air outside.

In December of that year, the cause of Legionnaires' disease was determined: a previously unrecognized bacterium that causes pneumonia. Further investigation determined that it was an airborne pathogen that was living in the hotel's air-conditioning system. Solving the mystery behind the eruption of Legionnaires' disease was truly one of the most remarkable accomplishments of modern-day epidemiology and microbiology.

Retrospective studies suggest that there had been at least two earlier, similar outbreaks. The first known epidemic occurred in the summer of 1957, in a Hormel meatpacking plant in Austin, Minnesota. In 1968, another outbreak occurred in Pontiac, Michigan. At the time, the name given to the then-unknown cause of the epidemic was *Pontiac fever*. For an unknown reason, these outbreaks were manifest by a much more benign course that usually resolved on its own without need for hospitalization.

Since 1976, there have been many small outbreaks of Legionnaires' disease throughout the Western world.[2] Most have been linked to aerosolized contaminated water dispersed from cooling towers, air-conditioning and plumbing systems, hot whirlpools and baths, respirator devices and nebulizers, or decorative water fountains. Outbreaks have begun in hotels, recreational facilities, and plumbing systems and showers in hospitals.

It also turns out that the bacterium that causes Legionnaires' disease can live in drinking water, especially unchlorinated water. Outbreaks are most common in the summer and when there has been recent flooding. (The risk of Legionnaires' disease increases when the weather is warm and humid.)

Risk factors for the illness include smoking, being older than fifty, having chronic lung disease, and having impaired immunity.

In the United States, Legionnaires' disease accounts for 2 to 9 percent of all pneumonia cases. An estimated eight to eighteen thousand people with Legionnaires' disease are hospitalized each year—but most experts believe these numbers are gross underestimates.

The fatality rate for Legionnaires' disease ranges from 5 to 30 percent. But for hospital-acquired Legionnaires' disease, the mortality rate is 28 to 50 percent.

Between 1995 and 2005, over thirty-two thousand cases of Legionnaires' disease—and more than six hundred outbreaks—were reported to the European Working Group for Legionella Infections. Many of these were tied to certain hotels in the Mediterranean region. The world's largest outbreak of Legionnaires' disease occurred in July 2001, when patients started showing up at a hospital in Murcia, Spain. Ultimately, 449 cases were confirmed, and at least sixteen thousand people were exposed to the bacterium.

The risk of Legionnaires' disease has grown steadily over the years, with a 192 percent increase in cases between 2000 and 2009. In 2015 alone, outbreaks were reported in a South Bronx hotel (128 cases), the Illinois Veterans Home in Quincy (46 cases), and San Quentin State Prison in Northern California (56 cases). In Flint, Michigan, a city already beleaguered by water contaminated with lead, at least eighty-seven people were sickened by Legionnaires' disease between June 2014 and October 2015. The sources of these infections is unknown.

In September 2016, an outbreak of twenty-three cases of Legionnaires' disease was reported in Hopkins, Minnesota, only a few miles from my home in Minneapolis. Within a month, the source of the outbreak was discovered—a cooling tower teeming with *Legionella*. And one year later, Disneyland shut down two cooling towers after people who visited the theme park came down with Legionnaires' disease.

Reports of new outbreaks of Legionnaires' disease throughout the United States pop up frequently. According to the CDC, the annual number of cases has increased more than fivefold since 2000. Why is this? More awareness and testing could be factors. But other contributors are thought to be an aging population and climate change.

The Enemy, Its Targets, and the Aftermath

At first, the July 1976 outbreak in Philadelphia baffled scientists, who were unable to identify its cause. Frustration was so great that congressional hearings were held that November. Fortunately, the CDC involved Dr. Joseph McDade, an expert on bacteria called rickettsiae, which grow inside cells. On December 28, 1976, Dr. McDade discovered a previously unknown bacterium that became known as *L. pneumophila* and that has the ability to live and reproduce in lung macrophages.

L. pneumophila obviously likes (and requires) water. But its favorite habitat is *inside other microbes*. It lives in a symbiotic relationship within water-loving protozoans, including amoeba. *L. pneumophila* hunkers down inside protozoa, beneath layers of a biofilm that stick to the surfaces of various kinds of plumbing, like water pipes and showerheads. While the precise mechanism whereby the biofilm is created is not understood, it appears that cohabitation of *Legionella* bacteria inside amoeba is involved. This intimate relationship serves the bacterium well as the biofilm not only conceals it from attack by cells of the immune system but protects against heat. (Water pipes crawl with millions of bacteria. Catherine Paul of Lund University in Sweden studies these ecosystems and has found at least two thousand different bacterial species living in water pipes. Fortunately, almost all are harmless; some may even help purify the water.)

The genus *Legionella* is now known to consist of fifty-eight different species of bacteria, of which at least six—*L. pneumophila*, *L. longbeachae*, *L. micdadei*, *L. feeleii*, and *L. anisa*—can cause Legionnaires' disease. The first of these is the most common cause of the illness.

Although most *Legionella* bacteria live in liquid environments, some species, such as *Legionella longbeachae*, are also found in moist garden soil. This species seems to be a significant cause of Legionnaires' disease in Australia.

Legionella generally likes things hot. It can withstand temperatures of up to 122 degrees for several hours, and it doesn't multiply at temperatures below 68 degrees. It grows optimally at between 97 and 98 degrees—essentially, the body temperature of human beings. After an incubation period of two to fourteen days, Legionnaires' disease causes symptoms such as fever, chills, muscle or joint pain, weakness, and loss of appetite. About half of infected people develop a cough that produces phlegm. Some develop a sharp chest pain that worsens with deep breathing or coughing. Gastrointestinal symptoms—diarrhea, nausea, vomiting, and abdominal pain—are also common, as are neurological manifestations such as headache, an altered state of consciousness, and seizures.

The chest X-rays and white blood cell counts of people with Legionnaires' disease are usually indistinguishable from people with more typical forms of pneumonia. However, there is now a urine test that can detect the bacterium. It can also be identified by culturing a person's phlegm.

Treatment and Prevention

Because *Legionella* are bacteria, they can be treated effectively with antibiotics. But penicillin and similar antibiotics don't work especially well because they don't get inside cells such as pulmonary macrophages in the lungs. Currently, either azithromycin or levofloxacin is recommended. Both have the ability to penetrate cells. An extended course of treatment is recommended for patients who already have some other illness, or whose immune systems have been compromised. During outbreaks, people who live in the affected area and who have a high risk of contracting the illness are often given prophylactic antibiotics, even if they exhibit no symptoms.

A vaccine for Legionnaires' disease doesn't yet exist, so the mainstay of prevention is cleaning up places where the bacterium likes to hang out and grow. Unfortunately, this isn't so easy, given the ability of *Legionella* to withstand relatively high temperatures and to live beneath a protective, tenacious layer of biofilm.

Lessons for the Future

"For many of us, water simply flows from a faucet, and we think little about it beyond this point of contact. We have lost a sense of respect for the wild river, for the complex workings of a wetland, for the intricate web of life that water supports."—Sandra Postel, director and founder of the Global Water Policy Project

We often take for granted the water we drink and bathe in, and we give little thought to the incredible microbial ecosystems that water supports—until something goes wrong, such as the emergence of Legionnaires' disease.

An especially vexing problem is the eradication of biofilms in pipes—particularly in hospitals, where there are so many high-risk people. Surveys have shown that *Legionella* species colonize hot water distribution systems in 12 to 70 percent of hospitals. New technologies are critically needed to prevent (or break up) biofilm formation in plumbing systems.

A key lesson of the Legionnaires' disease outbreak in Philadelphia is that air-conditioning systems and cooling towers could pose a threat to occupants of luxury hotels. Like many other emerging infections, this disease is a product of modern technology.

Another important take-home message from the discovery of *L. pneumophila*, a pathogen that was completely unknown prior to the Legionnaires' disease outbreak in 1976, is how little we actually know about how and when new infectious diseases will emerge. (The same can be said for MERS-CoV.) For now, we need to continue to expect the unexpected.

11

MICROBES IN THE WOODS

"The woods are lovely, dark and deep. But I have promises to keep, and miles to go before I sleep."—Robert Frost

Lovely as the woods may be, they are also home to innumerable ticks, which in turn are home to some dangerous microbes.

Ticks and microbes probably weren't on Robert Frost's mind when, in 1922, he penned his famous poem "Stopping by Woods on a Snowy Evening." In fact, the microbes discussed in this chapter, which are harbored by ticks in the woods, weren't even recognized back then.

The world has 899 known species of ticks. Fortunately, very few of these species carry pathogens that infect humans. Indeed, ticks play an important ecological role as an essential food source for many animals.

This chapter focuses on species that belong to *Ixodes*, a genus of hard-bodied ticks.[1] These ticks do pose big problems for human beings. Eliminating *Ixodes* ticks would be a great blessing to *Homo sapiens*. Getting rid of one species in particular, *Ixodes scapularis*, would be an excellent place to start. *I. scapularis* can harbor six different microbes—including three bacterial pathogens, two viral ones, and one protozoan—each of which is responsible for a great deal of human misery. Three of these microbial predators are discussed in this chapter.

LYME DISEASE

> "Lyme disease ticks me off."—Jeanne Braha, project director, public
> engagement, American Association for the Advancement of Science

The Lyme Disease Epidemic

The syndrome known as Lyme disease wasn't recognized until 1975. That was when a team of researchers from Yale University and the Centers for Disease Control and Prevention (CDC) investigated a mysterious outbreak in Old Lyme, Connecticut. Two mothers were deeply concerned about an unusual cluster of cases of what was originally called juvenile rheumatoid arthritis in their children, as well as in others in nearby towns.

The researchers, including Drs. Allen Steere, David Snydman, and Stephen Malawista, slowly began to connect the dots. They identified the symptoms (arthritis, a characteristic skin rash called erythema migrans, neurological problems, and heart disease); the geographic distribution (originally in the Northeast, but eventually spread to all fifty states); the tick species responsible (*I. scapularis* on the East Coast and in the Midwest, and *Ixodes pacificus* west of the Rocky Mountains); and the animals that harbor the disease (the white-footed mouse and the white-tailed deer).

A big breakthrough came in 1980, when Willy Burgdorfer, a medical entomologist at the Rocky Mountain Biological Laboratory, discovered unusual spirochetes—a distinctive type of spiral-shaped bacteria—in ticks sent to him from Shelter Island, New York. A year later, these spirochetes were determined to be the cause of Lyme disease. This heretofore unknown bacterial species was named *Borrelia burgdorferi* in his honor.

A fuller picture of Lyme disease has unfolded over the past four decades. For example, it's now known that *B. burgdorferi* isn't new. In fact, Lyme disease has been present in America and abroad for thousands of years. The 2010 autopsy of "Otzi the Iceman," a 5,300-year-old mummy discovered in the Alps in 1991 (which now is housed in the South Tyrol Museum of Archaeology in Italy), revealed DNA of *B. burgdorferi* in his body. This makes Otzi the earliest known human to have contracted Lyme disease.

In most of the United States, Lyme disease is spread primarily by *I. scapularis*, the blacklegged tick. During each different stage of its development—as larva, nymph, and adult—the tick is harbored by a different animal species. In all three stages, however, the tick requires a blood meal. Adult ticks feed on deer—so in the United States, *I. scapularis* is usually called the deer tick rather than the blacklegged tick. (In Europe, Lyme disease is primarily spread by *Ixodes ricinus*, the sheep tick.)

Most of us have encountered adult ticks. But it's the difficult-to-see nymphal stage ticks, which are about the size of the period at the end of this sentence, that are the problem. These are the ticks that most commonly transmit *B. burgdorferi* to humans. The primary host of nymphal ticks is the white-footed mouse.

Each year, about thirty thousand new cases of Lyme disease are reported to the CDC. But most experts believe this is a gross underestimate; probably well over three hundred thousand people get infected in the United States each year. Not surprisingly, the risk of human infection is greatest in the late spring and summer, when people are most likely to be trooping around in the woods—and when nymphal ticks latch onto innocent passersby. (Although essentially blind, they have a keen sense of smell—the carbon dioxide you release when you exhale really turns them on.)

Ixodes ticks are found throughout the United States. However, 99 percent of all reported cases of Lyme disease have been confined to the East Coast, the Northeast, and the northern Midwest. As reported in an article in the *New England Journal of Medicine* in July 2018, "Tickborne Diseases—Confronting a Growing Threat," tick-borne diseases, those carried by *Ixodes* as well as by other kinds of ticks, are increasing at an alarming rate.

The Enemy, Its Targets, and the Aftermath

Lyme disease is caused by bacteria belonging to the genus *Borrelia*. There are about twenty different species of *Borrelia*, but only three cause Lyme disease: *B. burgdorferi* (mainly in North America but also in Europe), *B. afzelli* (in Europe and Asia), and *B. garnii* (also in Europe and Asia). In 2016, researchers at the Mayo Clinic discovered another bacterial species, *Borrelia mayonii*, which also causes Lyme disease.

After someone has been bitten by an infected tick, the disease incubates for one to two weeks before symptoms appear. (About 7 percent of infected people have no symptoms at all. This is far more likely to occur in Europe than elsewhere.) The classic sign of early infection is a circular, outwardly expanding rash called erythema migrans (EM), which occurs at the site of the tick bite. However, this rash develops in only 70 to 80 percent of infected people, and its absence can lead to a misdiagnosis. The lack of a rash does not mean that the disease is not progressing or that later symptoms will not appear. (Another skin eruption—a purplish lump on the earlobe, nipple, or scrotum called borrelial lymphocytoma—sometimes appears in patients in Europe.) Infected people can also experience flulike symptoms such as fever, headache, muscle pains, and fatigue.

Within a few days to a few weeks after the infection begins, spirochetes may travel through the bloodstream to other parts of the body. When the disease attacks the nervous system, as happens in 10 to 15 percent of infected people, neurological problems, called neuroborreliosis, develop. These can include facial palsy (loss of muscle tone on one or both sides of the face), meningitis (inflammation of the covering of the brain), or encephalitis (inflammation in the brain itself). Bacteria may also lodge in the heart's electrical conduction system, giving rise to abnormal heart rhythms.

Untreated or inadequately treated patients may go on to develop what is called late disseminated disease, which is characterized by severe and chronic symptoms that affect many parts of the body, including the brain, nerves, eyes, heart, and joints. Joint problems (known as Lyme arthritis), usually involving the knees, occur in less than 10 percent of patients. A chronic skin disorder called acrodermatitis chronica atrophicans can sometimes occur, primarily in elderly Europeans. Chronic brain disease (encephalomyelitis) can also occur. Symptoms may be progressive and can include worsening cognitive impairment, leg weakness, an awkward gait, bladder problems, and even the development of psychosis (in such cases, the encephalomyelitis appears to cause a psychiatric disorder).

While Lyme disease is responsible for a good deal of human suffering, it is rarely fatal. In those who do die suddenly, a type of abnormal heart rhythm known as atrioventricular block is often the culprit.

In short, *B. burgdorferi* presents the human immune system with serious challenges.

Treatment and Prevention

Fortunately, antibiotic treatment is effective in combating *B. burgdorferi*. Doxycycline, amoxicillin, or cefuroxime axetil, taken by mouth for fourteen or twenty-one days, all work equally well.

Because laboratory tests aren't that helpful in establishing a diagnosis of Lyme disease, especially early on, treatment should begin as soon as a clinical diagnosis suggests Lyme disease. Most commonly, this means someone who was in the woods during tick season, in an area where Lyme disease is present, and has a suggestive skin rash (EM) and flulike symptoms. Because an EM rash isn't always present, flulike symptoms alone in a patient with known tick exposure can prompt treatment. A tick bite alone, on the other hand, isn't a reason for treatment.

The best way to prevent Lyme disease is to avoid tick-infested areas. Alternatively, when you go into the woods, use an insect repellent that contains DEET (in a concentration of at least 30 percent) and wear long pants and a long-sleeve shirt. When you get back home, examine your skin from head to toe, or have someone else examine you. If you see a tick, remove it carefully with tweezers.

A vaccine that prevents Lyme disease in humans (LYMEtrix) was approved by the FDA in 1998. Unfortunately, its entry into the market was thwarted by unanticipated complications and cost, and it was withdrawn. Currently, a Lyme disease vaccine is available for dogs, and new vaccines for humans are in development.

Lessons for the Future

"'Tis better to understand than to be understood."—Saint Francis of Assisi

Lyme disease continues to present science with several knotty problems. For starters, we need better diagnostic tests and an effective vaccine.

But there's an even more urgent need to understand how to manage the 10 to 20 percent of patients who receive treatment yet continue to complain of lingering fatigue, musculoskeletal pain, disrupted sleep, and cognitive difficulties. (This is posttreatment Lyme disease syndrome.)[2] Then there is the even larger group of patients who suffer with the disorder called chronic Lyme disease.[3] Clearly, there is a great need to unravel the mysteries behind these illnesses and come up with a treatment.

HUMAN GRANULOCYTIC ANAPLASMOSIS

"In the field of observation, chance only favors the prepared mind."— Louis Pasteur

The Human Granulocytic Anaplasmosis Epidemic

Human granulocytic anaplasmosis (HGA) is a tick-borne infection of white blood cells by the bacterium *Anaplasma phagocytophilum*. HGA is an emerging infection that is spread by the same insects as Lyme disease (*Ixodes* species of ticks). It is harbored by the same animals—the white-footed mouse and the white-tailed deer—and it has spread to the same parts of the world (primarily the U.S. East Coast, Northeast, and northern Midwest but also parts of Europe and Asia).[4]

Over 90 percent of cases of HGA occur in New England, New York, New Jersey, Wisconsin, and Minnesota. But, unlike Lyme disease, which was first recognized in New England, HGA emerged in my neck of the woods: the northern Midwest.

The first patient with HGA was a man from Wisconsin who died in a hospital in Duluth, Minnesota, in 1990. During the terminal phase of his illness, clusters of small, previously unknown bacteria were seen in a sample of his blood, inside a type of white blood cell called granulocytes or neutrophils. This chance observation became the key to the recognition of HGA.

Over the next two years, thirteen more people from northwestern Wisconsin and eastern Minnesota became infected. All had similar clusters of bacteria, called morulae, in their neutrophils.

Beginning in the mid-1990s, the number of HGA infections increased exponentially. Between 1995 and 2012, a total of 10,152 cases were reported to the CDC. Currently, cases are also increasing throughout Europe, where *Ixodes ricinus* is the tick that carries the illness, and in several Asian countries, including China, Korea, and Japan.

Fortunately, like Lyme disease, HGA is rarely fatal. But it often creates very serious symptoms—so serious that half of all people with

HGA need to be hospitalized. So far, at least seven people have died from HGA infections. The elderly and people with a compromised immune system are most likely to have severe symptoms.

The Enemy, Its Targets, and the Aftermath

While human infections caused by *A. phagocytophilum* weren't recognized until the early 1990s, veterinarians have been familiar with this microbe for at least two centuries. In Europe, the animal version of the disease (called tick-borne fever) has been observed in sheep, cows, and other ruminants since the early 1800s. *A. phagocytophilum* also sickens dogs, cats, deer, and reindeer.

You'll recall from chapter 10 that a type of bacteria called rickettsiae lives and thrives *inside* the cells of other creatures. In this sense, they are like viruses—but, unlike viruses, they have their own metabolic machinery. *A. phagocytophilum* is a type of rickettsiae.

Here is what is so remarkable about this bacterium: neutrophils are a type of white blood cell that is especially adept at killing bacteria. But the *A. phagocytophilum* bacterium targets, disarms, lives, and thrives *inside* neutrophils. The precise mechanism behind this is not yet fully understood.

The great majority of people with HGA suffer severe headaches, muscle pains, and tiredness. Some also experience nausea, vomiting, diarrhea, or respiratory problems, especially coughing. Severe HGA can be life threatening and can include complications such as shock, lung disease, hemorrhaging, kidney failure, heart inflammation, and neurological problems. About 5 percent of HGA patients require treatment in an intensive care unit.

Some folks with HGA, however, have no symptoms at all. In fact, scientists estimate that, in areas where *A. phagocytophilum* thrives, 15 to 36 percent of people have been infected without knowing it.

When someone shows up in a doctor's office with fever and a history of tick bite or tick exposure, HGA will be suspected, especially if the patient lives in or has visited one of the areas where HGA infections are common. Several different blood test results can suggest the illness, including a low platelet count (known as thrombocytopenia), a low white cell count (known as leukopenia), or high levels of a chemical made in the liver called transaminase, which is found in blood. To formally confirm that someone has the disease, however, a pathologist needs to examine some of the patient's neutrophils under a microscope and find morulae, the bacterial clusters that are unique to the disease.

An overactive immune response is thought to play a role in the development of severe HGA. Also, improperly functioning neutrophils appear to permit opportunistic infections to take hold.

Treatment and Prevention

When HGA is strongly suspected, prompt antibiotic treatment is essential. As of this writing, doxycycline is the drug of choice. Typically, doxycycline clears up most or all symptoms quite quickly, usually within forty-eight hours.

Because doxycycline is also often used to treat Lyme disease, someone who might have either illness—or who might be unlucky enough to have both at once—is automatically treated for both at once. In areas where both HGA and Lyme disease are endemic—like Minnesota—I routinely recommend doxycycline (one of the safest antibiotics) in patients with unexplained fever during the spring and summer months, especially if a patient reports hiking or camping.

A vaccine against *A. phagocytophilum* doesn't yet exist. The only way to prevent HGA is to avoid ticks, especially if you live in one of the areas where HGA is common.

Lessons for the Future

> "Education is a progressive discovery of our own ignorance."—Will
> Durant, American writer, historian, and philosopher

Even though a lot has been learned about HGA since the first case was identified in 1990, many questions remain. Most notably, why do so many people with the infection have no symptoms, while in others the illness can be life threatening? We can describe its symptoms, track its spread, recognize when someone has it, and treat it successfully. But how and why it does what it does remain almost complete mysteries. Researchers are currently working to find answers.

HUMAN BABESIOSIS

"A hidden danger is seeping into our blood supply."—Dave Mosher, science and technology journalist

The Human Babesiosis Epidemic

All of the emerging infections you've read about so far are caused by viruses or bacteria. Now let's look at one of the protists—*Babesia*—which causes yet another tick-borne infection, human babesiosis.

Babesia is a genus in the phylum Apicomplexa, which also includes the protozoan parasites that cause malaria and cryptosporidiosis, a serious gastrointestinal ailment. (You'll recall my discussion of malaria from chapter 6. We'll look at cryptosporidiosis in chapter 13.)

Of the several *Babesia* species, *B. microti* is by far the most common cause of human babesiosis. It is transmitted through *Ixodes* ticks, and—just as with *B. burgdorferi* and *A. phagocytophilum*—it is harbored by the white-footed mouse and the white-tailed deer. It too lives inside cells—but inside red blood cells, not white ones.

Historical records suggest that there were human cases of babesiosis in France as early as 1910. However, the first case wasn't documented until half a century later, in a Croatian herdsman who had had his spleen removed. (It turns out that the spleen plays an important role in defending the body against human babesiosis.)

The first case of babesiosis in a patient with a normal immune system was identified in 1969 on Nantucket Island, off the coast of Massachusetts. The disease was initially referred to as Nantucket fever. Since the mid-1990s, it has spread across the Northeast and northern Midwest, and the number of infections has grown markedly. Because the CDC didn't begin tracking babesiosis until 2011, the total number of cases is unknown. However, between 2011 and 2013, 3,862 cases were reported. Ninety-five percent of all cases have been in Connecticut, Massachusetts, Minnesota, New Jersey, New York, Rhode Island, and Wisconsin, though the illness has shown up in twenty-two states.

While *B. microti* is the primary species causing human babesiosis in the United States, a small number of cases caused by other *Babesia* species have been identified in northern California, Washington State, Kentucky, and Missouri.

In Europe, *B. divergens* is the predominant species, and the disease is spread by the *I. ricinus* tick. *B. microti*–like organisms have caused human babesiosis in Japan and Taiwan, and a new *Babesia* agent has been identified in South Korea. Sporadic cases of babesiosis also have been reported in Africa, Australia, and South America.

Most cases of babesiosis occur in the spring through early fall—which is, of course, when people are most likely to be walking around in the woods.

But tick bites aren't the only way to catch babesiosis. The disease is also transmitted through the transfusion of blood containing infected red blood cells. Although this is quite rare, babesiosis is nevertheless the most common blood-transfusion-transmitted infection in the United States. Between 1979 and 2011, more than 160 cases of transfusion-transmitted babesiosis (TTB) were reported to the FDA. In twenty-eight of these people, the infection contributed to their deaths. (To put this in perspective, however, over fifteen *million* blood transfusions are given to people in the United States each year.)

Fortunately, an effective test for screening donor blood for *B. microti* was reported in 2016. Because a quarter of healthy adults who get infected have no symptoms when they show up to donate blood, there is no way to know if they harbor the pathogen. Blood donors in areas of the country where babesiosis is common are routinely questioned about tick exposure, but this has not proven very effective. Thus, this new blood screening tool should protect recipients of blood from coming down with babesiosis.

The Enemy, Its Targets, and the Aftermath

As you'll remember from chapter 2, we can thank an infection in cattle called anthrax for Robert Koch's pivotal discovery in 1875 that confirmed the germ theory of disease.

In the case of babesiosis, it was another cattle disease, called febrile hemoglobinuria, that led Victor Babes, a Hungarian pathologist and microbiologist, to discover (in 1888) the microorganisms inside red blood cells that cause the disease. Appropriately, the illness was named after him. Five years later, Theobald Smith and Frederick Kilborne identified ticks as the way the ailment was transmitted to Texas cattle. This was the first time an insect was recognized as a way for an infection to be transmitted to a vertebrate host.

Now let's fast-forward almost a century to the human babesiosis epidemic.[5] The life cycle of *Babesia* in blood is very reminiscent of *Plasmodium*, the protozoan that causes malaria. In fact, an important diagnostic test for both protozoan infections involves examining stained blood specimens under a microscope for certain characteristic ringlike forms.

Half of all children and about a quarter of all adults who are infected with human babesiosis have no symptoms. But if symptoms do appear, the illness can be deadly. Most people with symptoms become ill one to four weeks after being bitten by an infected tick (or from one week to six months after receiving a transfusion of contaminated blood). The illness usually begins with fatigue and fever (with temperatures as high as 105.6 degrees). Chills and sweats are common and may be accompanied by headache, muscle or joint pain, loss of appetite, nausea, or sudden mood swings. In some cases, the illness may also cause an enlarged spleen or liver. The disease usually lasts for one to two weeks, but residual fatigue may persist for months.

The severity of the symptoms depends largely on the person's immune system. Severe babesiosis requiring hospitalization is most common in patients who have had their spleens removed or whose immune system is otherwise compromised due to cancer, HIV, organ transplantation, or treatment with immunosuppressive drugs. Other higher-risk groups include newborns, people over fifty years old, and people with chronic heart, lung, or liver disease.

Serious complications, such as heart, kidney, or liver failure; severe lung disease; a ruptured spleen; or coma develop in about half of hospitalized patients. Babesiosis kills about 6 to 9 percent of people hospitalized with the illness; in people with a compromised immune system, the mortality rate is 20 percent.

Babesiosis is one of only a handful of infections where the spleen is of crucial importance. (The spleen contains cells called macrophages, which remove certain types of infected or damaged cells.) The overproduction of cytokines by the immune system—the cytokine storm I described in chapter 9—also appears to play a role in the development of severe babesiosis. This is the same phenomenon that can make certain types of flu so damaging or deadly.

Treatment and Prevention

For patients with mild to moderate babesiosis and a normal immune system, the best and most common treatment is a combination of two drugs, atovaquone and azithromycin, taken orally for seven to ten days. An older drug regimen, intravenous clindamycin plus quinine taken by mouth, is recommended for patients with severe disease. Although these drugs have more side effects than atovaquone and azithromycin, they appear to be more effective in treating more serious cases of babesisosis.

Treatment for severe babesiosis may also include replacing blood containing infected red blood cells with blood from a healthy donor. The decision to initiate this therapy is based on an assessment of the percentage of red blood cells that contain protozoa, the severity of damage to the patient's red blood cells, and the presence or absence of organ failure. Consultation with an infectious disease specialist and a hematologist is recommended for all patients with severe babesiosis.

People with babesiosis can simultaneously be infected with *B. burgdorferi*, *A. phagocytophilum*, or both. If a combined infection is established or likely, doxycycline is normally added to the treatment regimen.

As with other tick-borne diseases, a vaccine against *Babesia* doesn't yet exist. Thus, preventing tick bites is the only effective way to prevent babesiosis.

Lessons for the Future

> "Take calculated risks. That is quite different from being rash."—
> George S. Patton

Tick-borne diseases, like the mosquito-borne infection West Nile fever, underscore the crucial roles that researchers who study insects (entomologists) and animals (zoologists) play in solving many of the puzzles of emerging infectious diseases. Future progress toward reducing, if not eradicating, these diseases depends on their creative involvement in determining more effective strategies to deal with vector-borne diseases in general. But until this happens, we need to be vigilant about the risks of walking in the woods.

But just how risky is it to walk in the woods? The North American Bear Center informs us that black bears have killed only sixty-one people across North America since 1900. It observes that a human being is much more likely to be killed by a domestic dog, bees, or lightning. All of this is quite accurate. However, the Bear Center concludes, "One of the safest places a person can be is in the woods." After you've read this chapter, you realize, *not necessarily*.

No one can accurately predict the odds of dying from a tick bite in the woods where *B. microti* resides. Clearly, the chances are very, very small. But people who have an increased risk of severe babesiosis—those with no spleen or an otherwise compromised immune system—would be wise to avoid deciduous forests and the edges of woodlands and open areas, where ticks may abound.

The sheer number of infections that ticks can carry may explain the existence of entonophobia, an ostensibly irrational—but perhaps not actually so irrational—fear of ticks. But for most people, walking in the woods remains a risk worth taking. Just remember to lather up with DEET, wear long-sleeved shirts and pants, and, at the end of your walk, check your body carefully for any ticks that may have joined you.

12

WHAT'S IN THE BEEF?

"If beef is your idea of 'real food for real people,' you'd better live real close to a real good hospital."—Neal Barnard, founding president, Physicians Committee for Responsible Medicine

First, a disclaimer—I'm not a vegetarian. I enjoy steak (medium rare) and hamburger (well done) as much as anyone. Second, I don't worry that microbes could be contaminating my food—at least not when I'm eating in the developed world.

That's not to say, however, that food-borne diseases in the United States aren't a problem. The Centers for Disease Control and Prevention (CDC) estimates that, each year in the United States, about 9.4 million people get ill from thirty-one known food-borne germs. A wide variety of foods can be contaminated by a wide array of microbes. While beef is often among them, many other foods—including eggs, poultry, fruits, vegetables, and fish—are more commonly to blame. The familiar pathogen *Escherichia coli* 0157:H7 is on the list of microbes causing these outbreaks. But it is far outnumbered by other bacteria and viruses.

As you know, one focus of this book is on germs (mortal enemies) that cause emerging infections—by definition, infectious diseases that emerged or reemerged over the past fifty years. Two of the most interesting of these infections are highlighted in this chapter. Both are classic zoonotic infections, linked in this case to cattle.

The first of these infectious diseases, variant Creutzfeldt–Jakob disease, is an exceedingly rare but devastating neurodegenerative disease that emerged in the United Kingdom in 1996, in the wake of so-called mad cow disease. It is caused by a prion, a type of pathogen that is so weird that I decided not to even include it in the early chapters of this book. I'll discuss prions shortly.

The second disease, enterohemorrhagic *Escherichia coli* colitis, is caused by toxin-producing *E. coli* 0157:H7. It sprang into the national spotlight in 1993 during a large outbreak associated with undercooked beef patties served at seventy-three Jack in the Box restaurants.

VARIANT CREUTZFELDT–JAKOB DISEASE (VCJD)

"Conventional people are roused to fury by departure from convention, largely because they regard such departure as a criticism of themselves."—Bertrand Russell

The vCJD Epidemic

The epidemic of vCJD is one of the most astonishing and baffling of all the emerging infectious diseases. Not only was it caused by a highly controversial pathogen, but it *broke the species barrier* by jumping from cattle to humans. This was unprecedented for this kind of pathogen.

The cause of vCJD is a pathogenic protein, typically designated PrP^{Sc}. In 1982, neurologist and biochemist Stanley Prusiner hypothesized that similar proteins caused scrapie, a transmissible brain disease in sheep. This idea, which was considered scientific heresy at the time, set off a firestorm. His colleagues asked incredulously, How can a protein, without the help of nucleic acid (DNA or RNA), possibly reproduce?

But Prusiner turned out to be right. Vindication came in 1997 when he was awarded the Nobel Prize in Physiology or Medicine for his work on misfolded proteins called prions.[1] (Actually, even today some scientists remain skeptical of Prusiner's hypothesis. Nonetheless, the evidence strongly supports it.)

Back in the 1980s, many scientific authorities also incorrectly argued that a devastating epidemic of bovine spongiform encephalopathy (BSE) in cattle—that is, mad cow disease—in the United Kingdom posed no threat to humans. That argument ended in 1995 when the outbreak of vCJD began and was quickly linked to BSE.

The first case of BSE in the world developed in 1994 in a cow on a farm in Sussex. As more cases accumulated in England, it became clear that the brain disease was similar to scrapie in sheep. Evidence soon suggested that the disease was spread when meat and bone meal from infected cattle was fed to calves.

Ultimately, the BSE epidemic in the United Kingdom was catastrophic. Between 1986 and 1998, more than 180,000 cattle were infected, and 4.4 million were slaughtered during the eradication program. Not only was this an overwhelming loss of life, but the economic losses to the beef industry, and to many thousands of farmers, were enormous.

Beginning in 1995, the first cases of vCJD in human beings were reported in the UK. As of May 2018, about 260 cases (all fatal) were reported worldwide. Most of the cases (178) occurred in the UK, with the remainder mainly in France (27 cases) and other European countries. Four cases were reported in the United States, two in Canada.

Epidemiological and scientific evidence linked almost all of these cases of vCJD to consumption of cattle products contaminated with the agent of BSE. Three cases in the UK were associated with blood transfusions.

The Enemy, Its Targets, and the Aftermath

Prions such as the one that causes vCJD are really tiny—smaller than viruses. They are so small, in fact, that they can't be seen with an electron microscope. As already mentioned, they are composed entirely of abnormally folded proteins. Because they don't contain nucleic acids (DNA or RNA), they can't reproduce, but they replicate by stimulating normal cellular prion protein to refold into the pathologic form called PrP^{Sc}. Like viruses, prions don't appear in the Tree of Life, as they don't have their own metabolism.

Even though they can't be seen by a microscope, thereby not fitting the definition of germs used in this book, it seems likely that they will be detected someday with more advanced microscopy.

Like mad cow disease, vCJD is transmissible and is characterized by a spongy degeneration of the brain. As the name implies, vCJD is a variant of another rare though far more common form of encephalopathy in humans called sporadic CJD (sCJD). sCJD occurs in about one in a million people. It too is uniformly fatal.[2]

In comparison to sCJD, patients who became ill with vCJD were younger (their median age at death was twenty-eight, as opposed to sixty-eight for people with sCJD). They were also sick longer, for a median of fourteen months, as opposed to four and a half months.

Early in the illness, people usually experienced psychiatric symptoms—most commonly depression or anxiety, and in about a third of cases, unusual, persistent, and painful body sensations. As the illness progressed, neurological symptoms set in, including unsteadiness, difficulty walking, and involuntary movements. As patients neared death, they became completely immobile and mute.

A hallmark of this type of encephalopathy is the long period between the consumption of contaminated beef and the first appearance of symptoms—typically several *years* or longer. It is thought that consumption of contaminated beef products during the BSE epidemic—possibly as early as 1986—was the main risk factor for humans.

Given the roughly ten-year span between the epidemics of BSE and vCJD and the impossibility of knowing how many people consumed contaminated beef, some experts estimated that cases of vCJD would number in the thousands. While this fortunately turned out to be an overestimate, vCJD isn't gone from our planet.

Treatment and Prevention

While several drugs have been used on a case-by-case basis to attempt to treat vCJD (as well as sCJD), none has proved beneficial. As of this writing, only palliative treatment is available for this tragic disease.

Keeping contaminated beef products out of the market is the key to preventing vCJD. During the mad cow disease epidemic, the UK moved quickly to cull potentially infected cattle. Since 1989, several control and prevention measures were implemented by the European Union, as well as by authorities in North America and elsewhere. In the United States, so far only four cows with BSE have been identified—the latest in California in 2012.

Because vCJD has also been associated with blood transfusion, monitoring of the blood supply is also important. Some countries have prohibited donations of blood from people who have lived in countries with a high risk of BSE.

Lessons for the Future

"Always keep an open mind and a compassionate heart."—Phil Jackson, former coach of the Chicago Bulls

One of the most important lessons learned from the vCJD epidemic is the crucial role in science of questioning conventional wisdom. The discovery that prions—misfolded proteins—can be infectious agents boggled everyone's mind at first. (It still boggles mine.) It felt a bit like discovering that fabric can be woven not only from wool, cotton, and other plant and animal fibers but also from petroleum.

Following the initial research on encephalopathy in sheep, other forms of the disease were identified. A spongiform encephalopathy in New Guinea, called kuru, was found to be transmitted by the eating of contaminated human brain tissue by cannibals. And cases of what is called iatrogenic CJD were linked to the unwitting medical or surgical use of contaminated materials—for example, human growth hormone, dura mater grafts, and liver or corneal transplants.

One of the many unanswered questions in the vCJD epidemic was whether the victims of the disease shared a predisposing factor. A genetic abnormality underlies the development of a similar form of CJD. But so far, no clear genetic susceptibility has been identified in vCJD.

Some scientists have questioned whether the relatively young age of patients with vCJD might be due to a failure to recognize the disease in old people who suffer from dementia. This seems doubtful to me given the other characteristics of vCJD. However, a growing body of evidence suggests that misfolded proteins could be the cause of other neurodegenerative diseases, such as Alzheimer's disease and Parkinson's disease.

The U.S. Department of Agriculture maintains surveillance for BSE in American cattle. In July 2017, they announced detection of an atypical BSE in an Alabama cow—the nation's fifth case and its first since 2012. Fortunately, the animal never entered the slaughtering process and therefore posed no threat to human health.

In recent years, another form of fatal spongiform encephalopathy has captured much attention in the United States. This time prions are infecting deer, causing what is called chronic wasting disease (CWD). CWD has been found in deer, elk, reindeer, and moose in at least twenty-six states and three Canadian provinces. In my state, Minnesota, the Department of Natural Resources announced in December 2018 that it was expanding its annual deer-hunting season to two consecutive weekends in an attempt to limit the spread of CWD. Because of an uptick of cases of sCJD in the past fifteen years in humans, the CDC is investigating whether prions in deer may have jumped to humans, as they did in the case of BSE.

Finally, one of the most important voids in knowledge laid bare by the vCJD epidemic is the absence of an effective treatment. I've been involved in the care of several patients with sCJD, and I must say that no disease is crueler.

ENTEROHEMORRHAGIC *ESCHERICHIA COLI* (EHEC) COLITIS

"Most E. coli bacteria help us digest food, synthesize vitamins, and guard against dangerous organisms. E. coli 0157:H7, on the other hand, can release a powerful toxin—called a 'verotoxin' or a 'Shiga toxin'—that attacks the lining of the intestine."—Eric Schlosser, author of *Fast Food Nation*

The Epidemic of EHEC Colitis

As we saw in chapter 3, a healthy human gastrointestinal tract is inhabited by about two thousand different bacterial species. All of these are either harmless commensals or, as in the case of most *Escherichia coli*, mutualists that contribute to our health. Unless you are running back and forth to the bathroom as you read this chapter, it is highly unlikely that you harbor one of the few harmful *E. coli* strains, such as Shiga toxin–producing *E. coli* (STEC) 0157:H7.

E. coli 0157:H7 emerged quietly in 1975 and caused several outbreaks in the 1980s. One outbreak, in 1982, involved the consumption of undercooked hamburgers prepared by McDonald's restaurants in Oregon and Michigan. But it was the Jack in the Box outbreak in 1993 that captured public attention and made this exotic-sounding bacterium a household name.

The Jack in the Box outbreak was traced to undercooked beef patties served in seventy-three restaurants in California, Idaho, Washington, Nevada, Louisiana, and Texas. It caused about seven hundred illnesses and

171 hospitalizations. Of the forty-three children who were hospitalized, thirty-eight suffered serious kidney problems (twenty-one required dialysis), and four died.

Health inspectors traced the contamination to the restaurants' Monster Burger sandwich, which had been on a special promotion (using the slogan "So good it's scary"). Sadly, had the Jack in the Box fast-food chain followed Washington State laws requiring that burgers be cooked properly to completely kill *E. coli*, this tragic outbreak wouldn't have occurred.

Investigation by the CDC identified six slaughterhouses as the likely sources of the contaminated beef. Over two decades later, in a congressional hearing on food safety in 2006, Senator Richard Durbin described this outbreak as a "pivotal moment in the history of the beef industry." It also served as a wake-up call to many regulators, including the U.S. Food and Drug Administration.

The official name of the disease caused by *E. coli* 0157:H7 is enterohemorrhagic colitis (EHEC). As this name suggests, once inside the colon, this germ triggers development of bloody diarrhea. As if that wasn't serious enough, it can also rupture red blood cells and precipitate kidney failure, the so-called hemolytic uremic syndrome (HUS). (Uremia refers to a raised level of urea that is normally eliminated by the kidneys.) Up to 10 percent of patients with EHEC infection develop HUS, and 3 to 5 percent of those patients die. Children and the elderly are the most likely to contract HUS—and the most likely to die of it.

EHEC-related damage to the colon, red blood cells, and kidneys are all linked to Shiga toxins produced by *E. coli* 0157:H7.

Since the Jack in the Box EHEC outbreak, many additional strains of Shiga toxin–producing *E. coli* have emerged in the United States. The CDC estimates that 265,000 STEC infections occur each year in America, and that *E. coli* 0157:H7 causes over 36 percent of these infections. Persons of all ages are susceptible, but the elderly and children suffer the greatest consequences.

Americans love beef. On average, each American eats more than fifty pounds of beef per year. About half of this—more than two billion pounds—is ground beef. Because almost 30 percent of Americans sometimes eat ground beef that's raw or undercooked, it's not too surprising that beef remains a major vehicle for outbreaks of EHEC colitis.

It may be unfair, however, to beef too much about beef as a vehicle for STEC infections. Many other outbreaks have been associated with contaminated lettuce, sprouts, cabbage, cilantro, apple juice, drinking water, and even prepackaged cookie dough. In fact, while writing this chapter in November 2018, the CDC was investigating a large multistate outbreak of *E. coli* 0157:H7 linked to romaine lettuce grown in Arizona. Nearly two hundred people were sickened, and five died.

It may also be unfair to demonize only *E. coli* 0157:H7. In 2015, for example, a different strain of STEC caused two multistate outbreaks associated with Chipotle restaurants, and in 2016, General Mills recalled ten million pounds of flour because of a potential link to an EHEC outbreak caused by yet another STEC strain.

Furthermore, other types of *E. coli* have gone rogue. One strain, *E. coli* 0104:H4, classified as an enteroaggressive *E. coli* (EAEC), was responsible for an outbreak of colitis and hemolytic uremic syndrome that started in Germany in 2011. Ultimately, 3,950 people from at least nine countries were sickened; eight hundred developed HUS, and fifty-three died. This epidemic was traced to contaminated fenugreek sprouts.

And then there are the enterotoxigenic *E. coli* (ETEC) strains that you may have encountered in fecally contaminated food. ETEC strains are a major cause of traveler's diarrhea, but they don't harbor Shiga toxin–producing bacteriophages. (If you are one of the 20 to 50 percent of people who develop watery diarrhea when traveling in a developing country, ETEC would be high on the list of potential culprits.)

The Enemy, Its Targets, and the Aftermath

So where did *E. coli* 0157:H7 come from? Clearly, it isn't a new microbe. Studies by evolutionary microbiologists suggest that this pathogenic strain separated from a commensal *E. coli* ancestor some time ago.[3] We don't have much of an idea when this occurred, but it was more than four hundred and less than 4.5 million years ago.

As we saw earlier, the virulence of *E. coli* 0157:H7 comes primarily from its production of Shiga toxin. Shiga toxin genes are mobile genetic elements that are "donated" by viruses that infect bacteria. Thus an important evolutionary microbiology question is, When and how did *E. coli* 0157:H7, as well as other strains of STEC, become infected with bacteriophages that code for Shiga toxin? As of now, we have no idea. (Incidentally, the toxins are named after Kiyoshi Shiga, who first described the bacterial origin of dysentery caused by the bacterium *Shigella dysenteriae*, which is also named after him.)

After an incubation period of two to ten days, most patients with EHEC colitis develop acute (and often extremely bloody) diarrhea. Other symptoms include abdominal cramping and vomiting. Somewhat surprisingly, they may experience no fever, or perhaps only a mild one. Most people recover in about a week.

Although the *E. coli* bacterium itself doesn't enter the bloodstream, in about 4 percent of cases its Shiga toxin does. If HUS develops, it usually occurs after about a week of illness. This serious complication is often signaled by dark or tea-colored urine, reduced urine production, and paleness (due to anemia). The Shiga toxins can also target the nervous system and cause seizures, neurological damage, and strokes.

Treatment and Prevention

As for all forms of diarrheal illness, it's essential to drink plenty of fluids to prevent dehydration. If symptoms are severe, or if you have bloody stools, call your doctor.

EHEC is a bacterial infection—yet, surprisingly, people who contract it should *not* take antibiotics. Studies have shown that when antibiotics kill the bacteria, they can prompt the release of Shiga toxins and actually worsen the illness.

If HUS develops, little can be done aside from blood transfusions and dialysis. However, experimental approaches to binding or neutralizing Shiga toxins are being developed. So are studies of probiotics that promote the growth of commensal bacteria, or that inhibit toxin production by STEC.

Thus far, efforts to eradicate *E. coli* 0157:H7 have not succeeded. This very hardy microbe is resistant to acid, salt, and chlorine. It can withstand freezing and can live in fresh water, in seawater, and on countertops for days. And it takes as few as five bacteria to cause disease, compared to millions of bacteria for most other food-borne pathogens.

What can you do to prevent EHEC? Wash your hands before preparing food (and after diapering infants or being in contact with cows or other farm animals). Avoid eating high-risk foods: unpasteurized milk or juice and soft cheeses made from unpasteurized milk. Above all, don't eat undercooked ground beef. If you have any doubts at all, use a food thermometer to make sure that any burger or meat loaf you cook has an internal temperature of at least 160°F (72°C). (This practice is also legally required of restaurants—although a recent study by the CDC indicated that eight in ten restaurant managers said their workers don't always take a final temperature of hamburgers with a thermometer.[4])

A vaccine to prevent EHEC colitis in humans doesn't yet exist. However, in a recent trial, immunization of feedlot cattle with a trial vaccine significantly reduced the amount of *E. coli* 0157:H7 in their feces.

Food irradiation is another promising food safety technology that can eliminate *E. coli* 0157:H7, as well as other EHEC strains and other bacterial species that commonly cause food-borne diseases. The FDA has approved the irradiation of meat and poultry, as well as fresh fruits and vegetables, spices, and other foods. The safety of irradiation has been extensively studied for more than forty years. It reduces or eliminates microbes but doesn't affect the nutritional value or taste of food. And neither the food nor the people who eat it become radioactive. However, food irradiation isn't a substitute for good food-handling practices.

Despite the evidence of its benefits, irradiated foods aren't yet widely available. Low consumer demand appears to have stalled the widespread use of this food safety technology.

Lessons for the Future

"Sacred cows make the best hamburger."—Mark Twain

EHEC colitis, along with a plethora of other food-borne infections, underscore the need for comprehensive food safety programs throughout our food distribution system, from the barnyard to your dinner plate.

The U.S. Department of Agriculture works with the FDA to promote food irradiation where it is appropriate. The USDA also controls the use of the word *organic* on food labels. Currently, foods that have been irradiated, no matter how they were grown or produced, cannot be labeled as USDA-certified organic. Organic growers and key people in the organic food industry adamantly support this decision. Whole Foods Market also insists that irradiation is not compatible with organic food production. In my opinion, though, it is time to make hamburger out of this sacred cow of the organic food industry.

13

GUT REACTIONS

"It is still just unbelievable to us that diarrhea is one of the leading causes of child deaths in the world."—Melinda Gates

Diarrhea is still a really big problem. The World Health Organization estimated in 2017 that 1.7 billion cases of diarrheal disease afflict children globally per year, and that each year diarrhea kills 525,000 children under the age of five. This translates into over 1,400 children dying each day—nearly sixty deaths an hour. Unsafe drinking water and inadequate hygiene—a scourge for 2.3 billion people in the developing world—underlie about 90 percent of these deaths.

But diarrhea also grips a lot of folks living in the developed world. As I mentioned earlier, I served as a doctor in the American Indian Health Service in Santa Fe, New Mexico, in 1971–1973. It was during this period that I decided to pursue advanced training in the field of infectious diseases, as I found these diseases piqued not only my interest but that of all my health professional colleagues. One of my assignments was chief medical officer for one of the pueblos, where I saw patients in the clinic, usually more than one hundred of them, twice a week. I became very fond of these stalwart people. So when a young child from the pueblo died of diarrheal disease, related in no small part to the difficulty of getting transportation to the Santa Fe Indian Hospital, I was heartbroken.

According to Herbert Dupont, a leading authority on gastroenteritis, about 179 million cases of acute diarrhea (defined as passage of three or more unformed stools per day for up to two weeks) occur each year in the

United States.[1] Most of these cases are caused by food-borne or water-borne pathogens.

But there is a major difference in waterborne infections in the developed versus the developing world. In the developed world, most water-borne infections come from contaminated recreational water—swimming pools, hot tubs, interactive fountains, and beaches.

A very long list of bacterial, viral, and protozoan pathogens can cause gastroenteritis, which is sometimes called stomach flu, but is completely unrelated to any influenza virus. You've already read about two of the bacterial culprits: in chapter 6, *Vibrio cholera*, the cause of cholera, and in chapter 12, *Escherichia coli* 0157:H7, the cause of the emerging infection called enterohemorrhagic colitis.

This chapter highlights three other emerging pathogens that cause trouble in the gastrointestinal tract. Two of these infectious agents—a protozoan, *Cryptosporidium parvum*, and a virus, norovirus—can be picked up by ingesting contaminated water or food. Norovirus has the distinction of causing the largest number of cases of acute gastroenteritis. The third agent, the bacterium *Clostridioides difficile*,[2] is the most lethal cause of gastroenteritis in the United States. It is highly resistant to antibiotics, and it haunts hospitals and other healthcare settings where people sometimes forget to wash their hands.

CRYPTOSPORIDIOSIS

> "The rains and runoff of March carried more than mud to Milwaukee. They brought the seeds of catastrophe."—Robert D. Morris, author of *The Blue Death*

The Cryptosporidiosis Epidemic

It was a banner year for emerging pathogens in 1993. In the same year that *E. coli* 0157:H7 debuted in the national news, an equally obscure protozoan, *C. parvum*, catapulted into the public spotlight.

C. parvum contaminated the drinking water in Milwaukee, Wisconsin. Over the span of about two weeks in March and April, 403,000 of the 1.61 million residents in the Milwaukee area developed gastroenteritis caused by this parasite. That was a quarter of the entire population. At least 104 deaths were attributed to the outbreak—mostly people with a compromised immune system, such as AIDS patients and the elderly.

Public health experts in Wisconsin and from the Centers for Disease Control and Prevention are credited with quickly establishing that the outbreak was caused by *Cryptosporidium* oocysts that passed through the filtration system of one of the city's water treatment plants. (An oocyst is an extremely tiny, hardy, thick-walled spore containing the parasite that lives in the stools of infected humans or animals.) Milwaukee's water comes from Lake Michigan; the source of contamination was traced to an outlet from a sewage treatment plant that released effluent into the lake.

To this day, the 1993 cryptosporidiosis outbreak in Milwaukee remains the largest waterborne disease outbreak documented in the United States. Just as the Jack in the Box *E. coli* 0157:H7 epidemic served as a wake-up call to the beef industry, the Milwaukee cryptosporidiosis outbreak rang the alarm for regulators of drinking water quality and wastewater management.

The first case of human cryptosporidiosis to ever be described occurred in 1976—in a three-year-old girl from rural Tennessee who suffered from severe diarrhea for two weeks. The first waterborne outbreak of the illness, however, occurred in 1984 and was attributed to fecal contamination of a public artesian well in Texas.

At about this same time, along with other infectious disease specialists, I watched *C. parvum* torture many AIDS patients. The source of their infections was usually unknown. In these patients with a severely compromised immune system, diarrhea was chronic, debilitating, and unresponsive to all forms of treatment. Cryptosporidiosis also killed many of these AIDS patients.

Since then, *C. parvum* has emerged as one of the most common waterborne pathogens worldwide. In 2012, 8,008 cases were reported to the CDC; 5.3 percent of these were associated with a detected outbreak. And in the summer of 2016, several hundred people in Columbus, Ohio, and Maricopa County, Arizona, were sickened by *C. parvum* that they picked up while swimming in public pools. The people most likely to get sick are children aged one to four, followed by elderly adults aged eighty or older. Recent studies in China suggest that children in developing countries are also at high risk of *C. parvum* infection.

Contaminated water (both drinking water and recreational water) and food (including fruits, vegetables, and raw beef) are the most common vehicles for transmission. Humans acquire the parasite by eating or drinking *C. parvum* oocysts.

Millions of oocysts can live in the stools of one infected human or animal. Yet experimental infection of human volunteers at the University of Texas showed that diarrhea developed in 40 percent of subjects who received only ten oocysts. Other studies suggest that a single oocyst is sufficient to cause illness. The shedding of oocysts in stools begins when symptoms begin and can last for weeks after diarrhea stops.

Like most emerging infections, cryptosporidiosis is a zoonotic infection. Calves appear to be the main carrier of *C. parvum*. However, other species of *Cryptosporidium* can be picked up from other animals.

The Enemy, Its Targets, and the Aftermath

Cryptosporidium was discovered in 1907 in the intestines of mice by an American physician and parasitologist, Ernest Tyzzer. But its veterinary and medical significance wasn't appreciated for almost seventy years, when the first human case of cryptosporidiosis was recognized. There are many different species of *Cryptosporidium*, but *C. parvum* and *C. hominis* are the most common causes of human infections.

Like *Plasmodium* and *Babesia*—other protozoans mentioned earlier in this book—*Cryptosporidium* belongs to the Eukaryota domain of life and to the phylum Apicomplexa. The outer shell of its oocysts—the infectious form that is passed in the stool—ensures its survival for long periods outside the body, and also makes it resistant to chlorine. And *C. parvum* can also resist many other common disinfectants. Once oocysts have been ingested, they transform into another form of the parasite that attaches itself to epithelial cells in the small intestine.

The spectrum of cryptosporidiosis symptoms ranges very widely. Some infected people show no symptoms at all;[3] others die from profound, fatal diarrhea. The integrity of someone's immune system largely determines how severe the disease will become. HIV/AIDS, organ transplantation, and other causes of immune deficiency are the major risk factors for severe disease.

Symptoms of cryptosporidiosis begin, on average, one week after a person becomes infected. The most common symptom is watery diarrhea; stomach cramps, nausea, vomiting, fever, and dehydration can also occur. Symptoms last for up to two weeks in people with normal immune systems, but in patients with compromised immunity, diarrhea can continue on and off for many months. In immunocompromised patients, cryptosporidiosis may spread to other organs, including the liver, gall bladder, bile ducts, pancreas, and respiratory tract.

Treatment and Prevention

Unsurprisingly, washing your hands—before eating and preparing food, after touching animals or children in diapers—is the single most important step in preventing the disease.

Many—but not all—commercially available home water filters remove *Cryptosporidium*. Every immunocompromised person should read the CDC's "A Guide to Water Filters," on the CDC's website, for specific recommendations.

Fluid and electrolyte replacement are the mainstay of therapy. Antidiarrheal medicine may help slow down the diarrhea. Children, pregnant women, and immunocompromised patients should be closely monitored by their doctors.

An antiparasite drug called nitazoxanide has been approved by the FDA for treatment of patients with a healthy immune system; however, the effectiveness of this medicine in people with a compromised immune system is unclear.

To date, there is no commercially available vaccine that is effective against *C. parvum*.

Guidelines are available on the CDC's website for keeping water free of *Cryptosporidium* oocysts, and for preventing cryptosporidiosis in general, in childcare settings, and in people with a compromised immune system.

Lessons for the Future

> "Thousands have lived without love, not one without water."—W. H. Auden

Seventy-two percent of our planet is covered in water. But 97 percent of that water is salty ocean water that's not suitable for drinking. Fresh water is a precious resource. And fresh water is teeming with germs—almost four hundred thousand bacteria per teaspoon. (Yet that's less than 8 percent of the number of bacteria in a teaspoon of seawater.) Thankfully, almost all of these germs are harmless—and some are beneficial.

But when bad actors like *Legionella pneumophila* (see chapter 10) or the oocysts of *C. parvum* are in the water, humans can be in big trouble. Although the transmission of these waterborne pathogens is very different—aerosolized water droplets in the case of *L. pneumophila*, the drinking of contaminated water in the case of *Cryptosporidium*—both microbes have taught us some of the same lessons.

Coincidentally, both *Legionella* and *Cryptosporidium* were discovered to cause human disease in 1976. And since then, both pathogens were responsible for worldwide outbreaks of pneumonia (Legionnaire's disease) or gastroenteritis (cryptosporidiosis). In both instances, we were quickly reminded of the paramount importance of supporting and monitoring our water supply—through proper plumbing, water quality management, sewage management, and the work of public health professionals.

NOROVIRUS

> "Keep calm and wash your hands—frequently and with soap."—Me to
> my patients (and anyone else who will listen)

The Norovirus Epidemic

Norovirus was first recognized as a cause of gastroenteritis in an outbreak in Norwalk, Ohio. (The genus name *Norovirus* is derived from *Norwalk virus*.) Although the outbreak occurred in 1968, the virus itself was not discovered until 1972, by the legendary virologist Albert Kapikian.

Over the next several decades, noroviruses blossomed into the leading cause of severe gastroenteritis worldwide. Currently, in the United States and the United Kingdom, norovirus is the most common cause of gastroenteritis outbreaks. In the United States, it is responsible for an estimated twenty-one million cases every year. Globally, the virus affects around 271 million people and causes over two hundred thousand deaths per year, mostly among young children, the elderly, and people with a compromised immune system. Currently, researchers estimate that at least half of all outbreaks of gastroenteritis around the world are caused by noroviruses.[4]

Outbreaks often occur in closed or semiclosed settings, such as long-term care facilities, hospitals, schools, prisons, clubs, dormitories, and restaurants. Outbreaks on cruise ships receive a lot of bad press (about twenty outbreaks are reported each year on ships that dock at U.S. ports), but cruise ships are the site of only about 1 percent of all reported outbreaks. Also, outbreaks on cruise ships have become less common in recent years. The CDC recorded just ten outbreaks of gastrointestinal illness on cruise ships in 2018—the second lowest level since 2001.

If you watched the 2018 Winter Olympics in Pyeongchang, South Korea, you learned that many star athletes were sidelined by norovirus. By February 8, 128 cases were confirmed. One wonders how many medals would have been won by these competitors had they not been pooped out by this nasty virus.

So why is norovirus such a successful pathogen? Mainly because it's extremely contagious. It takes as few as eighteen viral particles to cause infection, and a teaspoon of feces from an infected person can contain about 450 *billion* viral particles.

Humans are the only living creature known to harbor noroviruses, although recent studies suggest that oysters and dogs may harbor them as well. Transmission occurs by three general routes: from person to person, through contaminated food, or via water.

Norovirus gastroenteritis is often referred to as winter vomiting disease. This is because the disease is more common in the winter, and vomiting is a prominent symptom. Canadian researchers recently found norovirus RNA circulating in the air of rooms and hallways of hospitalized patients, suggesting that vomiting facilitates airborne transmission.

Salad ingredients and shellfish are the foods most commonly implicated in food-borne outbreaks of norovirus. In investigations of outbreaks, 70 percent of the time the infection was traced back to infected food handlers. (One outbreak in 2015 involved more than five hundred people who fell ill after eating at Chipotle restaurants. On February 8, 2016, all of Chipotle's roughly two thousand U.S. restaurants closed while the company held an all-staff meeting to address food safety issues.)

Given these figures, it's not surprising that the global economic burden of norovirus infections is pegged at a staggering $60 billion per year. And that's considered by some experts to be a conservative estimate.

Noroviruses are also the single largest viral cause of outbreaks of gastroenteritis associated with recreational water. Other sources of water-borne outbreaks include water from wells, municipally supplied tap water, and even ice machines.

The Enemy, Its Targets, and the Aftermath

Noroviruses are a genetically diverse group of single-stranded RNA viruses belonging to the Caliciviridae family. Like other RNA viruses, noroviruses are simple creatures. They carry only nine protein-coding genes. Recent studies suggest that noroviruses clump together in vesicles—clusters covered by a protective membrane—and that this feature is one of the reasons they are so virulent.

Noroviruses are notoriously hard to kill. They can stay alive on food, kitchen surfaces, and utensils for up to two weeks. They can withstand freezing and heat up to 140°F (60°C). They can also resist many common disinfectants and hand sanitizers.

When a person becomes infected with norovirus, the virus replicates inside their small intestine. Up to 30 percent of infected people show no symptoms. For the other 70 percent, however, symptoms typically begin after an incubation period of twelve to forty-eight hours. These include nonbloody diarrhea, nausea, vomiting, and abdominal cramps. In some cases, only diarrhea or vomiting occurs. A low-grade fever or body aches may also appear. These symptoms often cause infected people to call the illness stomach flu, but in fact noroviruses and influenza viruses are totally unrelated.

Symptoms may be severe, but they usually resolve in one to three days. Only 10 percent of patients are sick enough to seek medical attention, though some may require hospitalization. Norovirus-associated deaths most commonly occur among the elderly—often the result of outbreaks in long-term care facilities.

Noroviruses are shed in stools for about four weeks after the onset of infection. People who are infected with the virus but who don't become sick can nevertheless shed the virus in their stools.

Exactly how the human body defends itself against noroviruses isn't yet understood. One potential mechanism of defense involves microbes in the healthy gut microbiome, which we looked at closely in chapter 3. In support of this notion, Julie Pfeiffer and Herbert Virgin proposed in a recent article in *Science* that control of noroviruses in the gut is a joint (or "transkingdom") affair, involving interactions among bacteria, archaea, fungi, viruses, and eukaryotes. Their hypothesis may explain why so many people with infections don't develop symptoms—not only with noroviruses but with many other intestinal pathogens as well.

Treatment and Prevention

There is no specific medicine to treat norovirus. Because the disease is a viral infection, people who have it shouldn't be treated with antibiotics. Therapy focuses on preventing dehydration from fluid loss caused by vomiting and diarrhea. Medicines that control these symptoms may be helpful.

The CDC provides updated, evidence-based guidelines on prevention and control of norovirus gastroenteritis outbreaks, both in general and in healthcare settings. In practice, this often means removing or isolating infected people in long-term care facilities, hospitals, dormitories, and ships.

Hand washing with soap and running water for at least twenty seconds is an effective method for reducing the transmission of norovirus. Notably, hand sanitizers (gels, foams, and liquid solutions) appear to be less effective than hand washing with soap and water. In addition to washing your hands after every bathroom visit, disinfect food preparation equipment and surface areas with soapy water after preparing each food item and before you move on to the next food. Thoroughly wash fresh fruits and vegetables before eating them, and cook all meats, fish, and poultry thoroughly.

Surfaces can be sanitized with a solution containing household bleach. After first washing the surface with soap and warm, clean water to remove debris, next sanitize it with bleach. It is critical to read and follow the safety instructions on any household product you use.

Currently, there is no vaccine that prevents norovirus gastroenteritis. The good news, though, is that clinical trials of promising vaccines are underway.

Norovirus is similar in some ways to rotavirus, another emerging RNA viral pathogen that is transmitted by the fecal–oral route. Fortunately, rotavirus went fairly quickly from emergence to decline. The virus was discovered in 1973 and soon thereafter was recognized as the most common cause of diarrhea in children hospitalized in the United States. However, after a rotavirus vaccine was developed and introduced in 2006, the rate of infection in the United States dropped by more than 75 percent. This is a stellar example of the benefits of a successful vaccination effort. And due to financial support by nonprofit organizations and governmental agencies, rotavirus vaccines are now available in developing countries where the disease is still widespread.

Today, norovirus is the leading cause of acute gastroenteritis across all age groups. However, the development and success of the rotavirus vaccine suggests that we may soon develop a parallel response to norovirus.

Lessons for the Future

"The study of geriatrics begins in pediatrics."—Anonymous

Although the Norwalk species of norovirus was discovered about a half century ago, noroviruses have likely been with us for a very long time. The technology to identify the virus was available well beforehand, but it probably took a large outbreak, like the one in Norwalk, Ohio, to prompt researchers to turn their attention to it. No one knows how or where the virus originated.

Unlike most emerging infections, norovirus gastroenteritis isn't transmitted from animals to humans. But given the massive number of viral particles that float around our environment, it remains possible that some animal or animals harbor the virus, and the search for such an animal reservoir goes on.

As mentioned earlier, some evidence suggests that oysters may harbor the microbe. A recent study by researchers at the University of Helsinki's Department of Food Hygiene and Environmental Health also suggests that pet dogs harbor one form of norovirus. It is unclear, however, whether their findings point to potential transmission from dogs to humans, or from humans to dogs, in a process known as zooarthroponosis.

While reading this book, you've probably asked yourself more than once, What is it about young children and the elderly that places them at such high risk of infections and death? It's an important question—and with many infectious diseases, including norovirus gastroenteritis, the answer can sometimes make the difference between life and death.

I worked as an internist and infectious disease consultant at an inner-city hospital for more than twenty years, where I carried out clinical research on infections in the elderly. Throughout that time, this question was often on my mind. The best explanation appears to be that the immune system of some elderly people mirrors that of young children.

You'll recall from chapter 4 that one aspect of immunity, called innate immunity, is ready to defend you from the get-go when you're challenged by pathogens you haven't encountered. But a second aspect of immunity, referred to as adaptive immunity, must be built. This building process occurs every time you're challenged by a new pathogen. The cells involved are called T and B lymphocytes. And once they are primed by a first encounter, they spring into action (that is, they recognize) if they meet a pathogen again.

So, when you were a newborn, you had a lot of building to do; consequently, you were vulnerable to many infectious agents. And like most things, as we age our adaptive immune system isn't what it used to be. This phenomenon of waning immunity in some elderly people is often referred to as immunosenescence. In our seventies and beyond, the immune system becomes more childlike. And the susceptibility to certain pathogens that was first noticed in early childhood returns once again. This is also why many vaccines don't work as well in older adults as in young people.

But the good news is that this is not the fate of everyone. Many people who are in their seventies, eighties, and nineties (and some who are over one hundred) have an immune system that is as robust as those of middle-aged adults. Why are they so lucky? Nobody knows; it is one of the many mysteries of immunity.

The bad news for society is what is sometimes called the silver tsunami. By the year 2050, people over sixty-five will represent more than 20 percent of the U.S. population. This means that more people will be living in long-term care facilities—and more infections such as norovirus gastroenteritis will likely result.

CLOSTRIDIOIDES DIFFICILE INFECTION

"In the past five years, C. diff has spread across the globe, helped in large part by air travel, the availability and frequent use of antibiotics, and the graying of the world's population."—J. Thomas LaMont, author of *C. diff in 30 Minutes*

The Epidemic of *Clostridioides difficile* Infection

In 1977, just as I finished my training as an infectious disease specialist, I was asked to see a patient whom I will call Mr. Oturo. He was a sixty-eight-year-old retired schoolteacher with life-threatening diarrhea. My colleagues and I did everything we could to restore his health—but, to our shock, within five days he died.

Mr. Oturo was one of the first patients I cared for who had what would later become known as an emerging infection. It took us some time after his death to discover that the disease that had killed him was caused by a bacterium called *Clostridioides difficile*. (It was named *Clostridium difficile* at the time.) The disease, called pseudomembranous colitis, now kills close to thirty thousand Americans per year.

Back then, none of us would have believed that, someday, the treatment for this disease would entail the use of human feces, in what is known as fecal microbiota transplantation.

Also in those early days of my career, I knew nothing about Carl Woese, the new domain of microorganisms that he had named Archaea, or the technology that he and his colleagues invented. As we saw in chapter 1, that technology led to a new understanding of the Tree of Life and to much of what we know today about the human microbiome. With this technology, researchers later learned that the gut, where *C. difficile* was wreaking havoc in Mr. Oturo, is populated by untold trillions of microbes, some of which fend off disease-causing germs. Had we known what we know now about the human microbiome, we probably would have been able to save Mr. Oturo's life.

In the early days of the pseudomembranous colitis epidemic, a link between antibiotics, the emergence of *C. difficile*, and the development of colitis (inflammation of the colon) was quickly established. We clearly saw that the only patients who came down with *C. difficile* infection (CDI) had taken an antibiotic prior to its development. At the time, the antibiotic clindamycin was widely used in American hospitals, and it seemed that the problem was uniquely associated with this antibiotic. In fact, this was the antibiotic that Mr. Oturo had been treated with. Researchers later learned that clindamycin, along with many other antibiotics, actually eliminated from the gut the very bacteria that held *C. difficile* at bay.

Now let's fast-forward about four decades to the present. By now, essentially *all* antibiotics are known triggers of CDI, which has become the leading cause of hospital-acquired diarrhea in North America and Europe. The number of cases of CDI has skyrocketed, to about five hundred thousand cases per year in the United States alone. Much worse, the mortality rate has increased *twentyfold*. Annually, fifteen thousand patients hospitalized in the United States with CDI die. Hospital-acquired CDI is also very expensive to treat, more than quadrupling the cost of many people's hospitalization.

To make matters worse, CDI recently moved beyond hospitals into our cities and towns, where about 30 percent of cases are now acquired. CDI has become *the* leading cause of fatal gastroenteritis in the developed world, by far.

In some ways, understanding the reasons for *C. difficile*'s monumental success as a pathogen is simple. It is transmitted from person to person by a fecal–oral route, usually by the contaminated hands of healthcare providers. No food, no water, and no animal reservoir is necessary. And there are only two groups at risk. But those two groups include lots of people.

First, anybody who takes an antibiotic is at risk of contracting CDI. And where are many of the patients who receive antibiotics concentrated? Hospitals and long-term care facilities. Roughly half of all hospitalized patients, and many nursing home residents, are treated with antibiotics.

The second risk factor, reminiscent of norovirus infection, is being old. One out of every three healthcare-associated CDI cases involves someone sixty-five or older. More than 80 percent of the deaths from CDI occur in people in this age group. Recent studies at the University of Virginia, reported in the *Journal of Infectious Diseases* in 2018, suggest that the increased risk of CDI in elderly people may be associated with age-related alterations of their gut microbiomes.[5] Two additional factors that explain the mounting impact of *C. difficile* relate to its biology. Like its clostridial cousins, when *C. difficile* is stressed, it forms spores, and these spores are extraordinarily tough. They can survive in a dormant state for up to five months on surfaces. When they find themselves in an anaerobic environment—that is, one that lacks oxygen—the spores germinate. One of the most common and available anaerobic environments is the human colon.

In addition, to defend themselves against our microbial allies in the gut, as well as against cells of the immune system, *C. difficile* bacilli produce toxins that damage the colon. In this battle between the good germs (the intimate friends in our gut microbiome) and bad germs (toxin-producing *C. difficile*), our colon gets caught in the middle.

The CDI epidemic has swept over us in two waves. Before the twenty-first century, during the first wave, CDI was considered a serious but manageable problem. But beginning in 2001, the rates of CDI infection began skyrocketing, first in the United States, then in Canada and many European countries. At the same time, more and more cases required emergency colon surgery, and the death rate from the illness increased. These alarming developments were due to the emergence of a very harmful *C. difficile* strain, designated as BI/NAP1/027.

The Enemy, Its Targets, and the Aftermath

C. difficile was first isolated from the stool of healthy newborns by Ivan Hall and Elizabeth O'Toole in 1935. Its name reflects how difficult it is to grow the microbe in a laboratory culture. These investigators also demonstrated that the bacterium produces a toxin that is highly lethal to mice. (Incidentally, while its genus name *Clostridium* recently was changed by taxonomists to *Clostridioides*, its clinical moniker, *C. diff*, is likely to stick.)

Over forty years elapsed, however, before two teams of researchers (W. L. George and John Bartlett and their colleagues) established the link between *C. difficile*, antibiotic treatment, and pseudomembranous colitis.

Much research attention has focused on the two toxins that *C. difficile* produces, named (rather uncreatively) toxin A and toxin B. Of the two, toxin B is the most potent at inducing severe inflammation in the colon. The newly emerged BI/NAP1/027 strain, referred to more simply as NAP1, is hypervirulent. This means this strain produces an exceptional amount of toxins that are more potent, and BI/NAP1/027 is more readily transmitted than less virulent strains. It is also a hyperproducer of spores.

In his article "Host-Pathogen Interactions in *Clostridium difficile* Infection: It Takes Two to Tango," David Arnoff points to the convergence of NAP1 strains and the growing number of elderly hospitalized patients to explain the dramatic increase in mortality of CDI.

People who acquire CDI in the community, rather than in a healthcare institution, are generally younger and are much less likely to get seriously ill or die. Nevertheless, 40 percent of these people require hospitalization.

Exactly where and how people pick up toxin-producing *C. difficile* in the community is unclear. Bacteria belonging to the genus *Clostridium* are everywhere in nature. Potential sources of spores include soil, water, pets, meats, and vegetables. Whether *C. difficile* is joining its relatives in these environs is unknown (and if so, unnerving).

Our body's immune defense against *C. difficile* in the gastrointestinal tract is complex, involving a delicate balance of the gut microbiome and cells of the innate and adaptive immune systems. Recently, researchers at the University of Michigan in Ann Arbor reported that no single microbial species defends the colon against *C. difficile* gaining a foothold. Rather, resistance was associated with a cooperative interaction of five different groups of the colonic microbiota. [6]

Antibodies to toxin B are known to protect the gut against CDI. And recent studies by researchers at the University of Virginia suggest that the toxin damages eosinophils, a type of white blood cell that also protects the gut against *C. difficile*.

In hindsight, the initial discovery by Hall and O'Toole of *C. difficile* in the stools of healthy newborns isn't surprising. In early infancy, most children carry strains of *C. difficile* in their bodies—and the vast majority have no symptoms of the illness. *C. difficile* is also present in the colon of 2 to 5 percent of all adults who show no symptoms.

The key trigger to becoming ill is the disturbance of the gut microbiome by an antibiotic that eliminates friendly competitors. Illness begins with uncomplicated watery diarrhea (three or more loose stools in twenty-four hours). This early symptom of CDI is indistinguishable from a more or less benign illness called antibiotic-associated diarrhea (AAD), which occurs in 10 to 15 percent of patients treated with an antibiotic. Laboratory testing of fecal matter for *C. difficile* or its toxins helps sort out AAD from CDI.

When CDI progresses, however, massive pseudomembranous colitis develops. In some cases, marked expansion of the colon occurs—a condition known as toxic megacolon. Notably, in up to 20 percent of patients with advanced disease, the diarrhea subsides and is replaced with consti-

pation and a ballooning abdomen. Patients with severe CDI may experience low blood pressure, kidney or respiratory failure, and evidence of
damage throughout the body.

C. difficile itself usually stays put in the colon, provoking a hyperacti-
vated immune response that may require the emergency removal of the
colon (called a colectomy). While the overall mortality rate of CDI is
about 5 percent, the mortality rate following a colectomy is close to 70
percent.

Treatment and Prevention

Somewhat paradoxically, the first line of treatment of CDI is an antibiotic—either metronidazole or vancomycin, taken by mouth. Vancomycin
is preferred for more serious infections. Most patients respond favorably
to this treatment, but about 25 percent develop a worse form of the illness
known as recurrent CDI. As its name suggests, the illness becomes chronic—appearing, subsiding for a time, and then reappearing. If severe colitis develops, other antibiotic regimens and routes of administration
(intravenous or by enema) are tried. In people considered at risk of dying,
surgical removal of the colon is essential. However, before this pretermi-
nal stage of CDI is reached, fecal microbiota transplantation—a means of
delivering our intimate microbial friends to the site of infection—is now
highly recommended for recurrent CDI. (Much more on the topic of fecal
microbiota transplantation in chapter 16.)

Treating patients with recurrent CDI is very challenging. The first
recurrence is generally treated with vancomycin, often in combination
with other antibiotics that have proven effective for CDI. Treatment with
probiotics is an appealing approach. As is discussed in chapter 17, probiotics are defined as "live microorganisms which, when administered in
adequate amounts, confer a health benefit on the host." Although treatment and prevention of CDI by consumption of probiotics (nonpathogen-
ic bacteria or fungi) has been attempted, as of this writing insufficient
data is available to recommend them.

Unfortunately, a vaccine that prevents CDI isn't available. But the
success of some novel immunological treatments of recurrent CDI is
encouraging. One strategy uses antitoxins via a druglike molecule called
ebselen, which shuts down toxin production. Positive results of two randomized clinical trials of bezlotoxumab, a human monoclonal antibody to

C. difficile toxin B, were reported in January 2017. Yet another strategy involves administration of hyperimmune colostrum (a form of milk generated in late pregnancy) obtained from cows that have been successfully immunized against toxins A and B.

Prevention is, of course, much preferred to infection and treatment. Because *C. difficile* is transmitted from person to person via contaminated hands, thorough hand washing and the use of gloves are the primary forms of prevention.

You might think that all healthcare providers pay strict attention to hand hygiene. Sadly, this is not the case. Time and again, doctors, nurses, and others who come in contact with patients fail to properly wash their hands or wear gloves. This isn't because they don't care or don't believe in the germ theory of disease. It's because they're very busy and sometimes forget.

In May 2016, the CDC launched its Clean Hands Count campaign. Time will tell whether this infection control program is more effective than previous efforts to encourage hand hygiene. (This campaign is also highly relevant to the prevention of other hospital-associated emerging infections, as you'll discover in chapters 14 and 15.)

But even with perfect hand hygiene, *C. difficile* won't give up easily. Its spores are acid and heat resistant, and they aren't killed by alcohol-based hand cleansers or routine surface cleaning.

Patients with CDI are routinely isolated from other patients to help prevent the transmission of infection. A recent Canadian study suggests that screening all patients for *C. difficile* on their admission to the hospital and isolating those who are carriers of the microbe can also reduce the incidence of CDI. And in 2016 the worrisome results of a study carried out in New York City were reported, indicating that patients are at increased risk of developing CDI if the prior occupants of their hospital beds had received antibiotics.

Another novel way to use good bacteria to prevent CDI takes advantage of *C. difficile*'s spore-forming capacity. In this case, however, the spores are from a non-toxin-producing *C. difficile* strain. Patients swallow these spores to establish competition in the gut, thereby preventing colonization by toxin-producing bacterial strains.

Results of a randomized placebo-controlled trial of this approach, published in the *Journal of the American Medical Association* in 2016, showed that patients with a history of treated, recurrent CDI who received spores from a non-toxin-producing strain had fewer recurrences of the illness.

Because antibiotics trigger the development of CDI, and antibiotic use is deemed unwarranted 50 percent of the time, promoting the judicious use of antibiotics, called antibiotic stewardship, is now recognized as a leading strategy to prevent CDI. To this end, the CDC has created a list of core elements of hospital antibiotic stewardship programs. In a parallel move, the Agency for Healthcare Research and Quality furnishes hospitals with a tool kit for implementing antibiotic stewardship programs to reduce CDI. Such programs help prevent the side effects of antibiotics, such as CDI.

Preliminary analyses of national data from 2011 to 2014 by the CDC suggested the incidence of CDI was decreasing. This encouraging finding was attributed to the implementation of antibiotic stewardship programs in hospitals across the country. But paradoxically, in July 2017, the results of a large retrospective study by researchers at the University of Pennsylvania disclosed an increase in recurrent CDI.

Antibiotic stewardship programs are now embraced by all hospitals and long-term care facilities in the United States. These programs address not only CDI but another very high-priority health problem: the emergence of antibiotic-resistant bacteria, which I'll discuss in chapters 14 and 15.

Lessons for the Future

"Sometimes if you want to see a change for the better, you have to take things into your own hands."—Clint Eastwood

Infections such as CDI and norovirus gastroenteritis, both of which are spread primarily by the fecal–oral route, have sparked a reawakening of strict attention to proper hand washing. To help prevent these infections, wash your hands with soap and water for at least twenty seconds after visiting the bathroom and before preparing food.

In this era of patient-centered care, doctors want your help in combating these and similar infections. If you are hospitalized, insist that everyone who walks into your room immediately wash their hands. If you are visiting someone else in the hospital, do so yourself—and if the person you are visiting doesn't make the same demand of whoever else enters the room, do so on their behalf. This may annoy people, but it may also save someone's life. In the hospital, scrub your hands for forty to sixty seconds with soap and water; alcohol-based hand sanitizers aren't strong enough. If the person in the room has CDI, everyone who comes in contact with them needs to wear gloves and a gown.

Your help is also needed in combating the overuse of antibiotics. Whenever one is prescribed for you or someone you care for (e.g., your child, an elderly relative, or someone who is too ill to think straight), ask why it was prescribed, whether it is genuinely necessary, and whether a form of treatment without an antibiotic might be equally effective.

Remember that antibiotics *only* work for bacterial infections. If you have a viral infection, such as norovirus gastroenteritis, or a fungal infection, such as candidiasis, or a parasitic infection such as malaria, antibiotics won't help. Furthermore, they can make matters much worse—for example, by leading to CDI.

CDI should provide everyone with a healthy respect for poop—partly because it can harbor pathogens, but mostly because poop is a by-product of friendly germs in the gut microbiome. (You'll read more about how poop prevents recurrent CDI in chapter 16.)

Should you now be thoroughly pooped from reading about diarrhea, there is one piece of really good news that I want to leave with you. A 2017 study published in the journal *Lancet* reported that diarrhea-related deaths declined about 20 percent from 1.6 million in 2005 to 1.3 million in 2015. While a reduced mortality wasn't seen in wealthy countries like the United States, where CDI is the main cause of diarrhea-related death, the number dropped dramatically in low-income countries, especially among children. For children under five, diarrheal diseases claimed the lives of 35 percent fewer children.

14

WHEN BEAUTY ISN'T SKIN DEEP

"The story of MRSA—methicillin-resistant *Staphylococcus aureus*, or drug-resistant staph—is the story of how we took the antibiotic miracle for granted, and how we failed to plan for the creative survival tactics of the bacteria that are mankind's oldest companions."—Maryn McKenna in *Superbug: The Fatal Menace of MRSA*

Few bacteria are more feared by physicians than *Staphylococcus aureus*, commonly called staph. It was one of the first pathogens discovered during the 1880s, and it has been recognized ever since as the most common cause of painful infections of the skin (cellulitis) and adjacent soft tissue (abscesses). When it invades the bloodstream (so-called blood poisoning or bacteremia), *S. aureus* can set up shop on heart valves (endocarditis), in bones (osteomyelitis), in lungs (pneumonia), or in the brain (meningitis or brain abscess). And in the pre-antibiotic era, these infections usually carried a death sentence.

Ironically, we have *S. aureus* to thank for the serendipitous discovery in 1928 of penicillin by Alexander Fleming. He observed that a substance produced by a *Penicillium* mold that accidentally floated onto a culture plate killed the staph on the plate. In 1945, Fleming shared the Nobel Prize for Physiology or Medicine along with Howard Florey and Ernst Chain, who were the first to use penicillin clinically in 1941.

While penicillin clearly deserves its status as a miracle drug, Fleming warned in 1945 that bacteria might be "educated to resist penicillin." Sure enough, by the late 1950s infections caused by penicillin-resistant *S. aureus* were common.

Fortunately, in 1959 chemists synthesized a related antibiotic called methicillin, which was impervious to the enzyme produced by *S. aureus* that inactivates penicillin. Not to be outmaneuvered, however, methicillin-resistant *S. aureus* (abbreviated MRSA and pronounced "mersa") began appearing in the 1960s. Because of its virulence and its resistance to many antibiotics, MRSA is considered one of the first superbugs. (I'll discuss superbugs in general in chapter 15).

In the 1980s, MRSA began terrorizing patients, first in hospitals and then, in the 1990s, out in the community. This chapter tells the story of the emergence of these two parallel epidemics, with a focus on skin and soft-tissue infections.

But before I discuss these epidemics, you need to know the formal definition of MRSA, which is isolates of *S. aureus* that are resistant to all currently available beta-lactam antibiotics. This includes penicillin, ampicillin, methicillin, and other penicillin-derived drugs. It also includes all the cephalosporins—a long list of antibiotics that once were tried-and-true agents for treating *S. aureus* infections.

A brief discussion of how bacteria like *S. aureus* develop antibiotic resistance is in order. The different mechanisms underlying antibiotic resistance are related to changes in the DNA of the bacteria.

As you'll recall from chapters 1 and 2, bacteria have been around for almost four billion years—and, along the way, they evolved genes for combating microbial competitors, either by producing antibiotics or by developing mechanisms to resist their activity. In some cases, evolution favored bacteria that were fit with a gene that codes for an enzyme that destroys the antibiotic. The best example of this is penicillinase, which destroys penicillin. This became a widespread problem with *S. aureus* beginning in the 1950s.

In the case of methicillin resistance, a gene called *mecA* is responsible. It encodes a novel protein in the cell wall of staph that prevents the binding of methicillin. As a result, methicillin and similar antibiotics don't work.

METHICILLIN-RESISTANT *STAPHYLOCOCCUS AUREUS* (MRSA)

"MRSA is in every hospital in the United States, just lurking there."—
Lisa McGiffert, director, Consumers Union Safe Patient Project

The Epidemics of MRSA Skin and Soft-Tissue Infections

Soon after methicillin was enthusiastically welcomed as a treatment of penicillin-resistant *S. aureus*, MRSA strains were identified in 1961 in the United Kingdom. And in 1968, the first hospital outbreak of MRSA was reported in Boston.

When I completed my training in infectious diseases about a decade later, such infections—called healthcare-associated MRSA (HA-MRSA)—were oddities. Only a few hospitals in the United States were battling MRSA, in their burn or dialysis units. Similarly, in Europe, some (but not all) countries reported HA-MRSA infections. And with the exception of a small cluster of intravenous drug users in Detroit in the 1980s, community-associated MRSA (CA-MRSA) was unheard of.

These were the early days of the two parallel epidemics of MRSA infections. Then, in the 1980s, in rapid succession, I witnessed the expansion of HA-MRSA. It soon became the norm rather than the exception in all hospitals in the United States, as well as in many other countries. After consulting in 1998 on one of the first cases of CA-MRSA in the United States—a previously healthy seven-year-old girl with a hip infection that eventually killed her—I witnessed the alarming spread of CA-MRSA to other groups of people who had no connection with hospitals.

In most, but not all, cities in the United States, CA-MRSA is now the most common pathogen cultured from patients with skin and soft-tissue infections in emergency departments. According to the Centers for Disease Control and Prevention, in 2005 about ninety thousand invasive MRSA infections and twenty thousand deaths occurred in the United States. But these figures include only infections of the bloodstream or internal body sites. They don't take into account the large number of CA-MRSA infections involving the skin or soft tissues.

What's behind both MSRA epidemics? Some of the same factors you read about in the last chapter on *Clostridioides difficile* infection (CDI). Like *C. difficile*, MRSA is commonly transmitted from person to person by the contaminated hands of healthcare providers. This is especially true in hospitals, where MRSA now has a very firm foothold. *S. aureus* infections are the bane of hospitals.

In patients with CA-MRSA, contact with contaminated inanimate objects seems to play an important role in transmission. In a recent survey of randomly chosen homes in New York City, MRSA was found as an environmental contaminant in 20 percent of all households.

Complicating matters is the fact that *S. aureus* colonizes the nostrils—a moist body site—of about one-third of all healthy people. Usually this is methicillin-sensitive *S. aureus* (MSSA), the penicillin-resistant strain that can be easily killed by methicillin or something similar. Only about 2 to 7 percent of healthy people harbor the deadlier MRSA in their noses. Nevertheless, when someone does come down with symptoms of MRSA, it is often difficult to know whether a patient's nose or some surface in their environment was the source of infection.

While the epidemics of HA-MRSA and CA-MRSA share some features, there are also some striking differences.[1] For one thing, infections involving the skin (cellulitis) or soft tissue (abscesses) are far more common with the community version (CA-MRSA) than with the hospital version (HA-MRSA). The main skin and soft-tissue infections caused by HA-MRSA appear in postoperative wounds. Most feared in hospitals—and most dangerous—are pneumonia and bloodstream infections (bacteremia). The latter type of infection is often the result of a contaminated intravenous device.

Another distinguishing feature of CA-MRSA versus HA-MRSA infections is their risk groups. By definition, people at risk of HA-MRSA are hospitalized. They often are elderly, have undergone surgery, or have chronic underlying medical conditions. In many cases, they have catheters, intravenous needles, or other devices inserted into their bodies. In contrast, a patient with CA-MRSA can be anyone.

Many reports suggest that MRSA is easily transmitted in any setting where people are in close contact, including households, day care centers, military installations, prisons, locker rooms, and so on. MRSA has been dogging college and professional sports teams for more than a decade. Athletes who get a lot of breaks in the skin from abrasions, such as football players, are particularly prone to CA-MRSA. MRSA has sidelined or ended the careers of a long list of famous professional football players. (As just one example, Lawrence Tynes, a star placekicker for the

Tampa Bay Buccaneers, ended his football career in 2014 because of a losing battle against a MRSA knee infection. And he wasn't alone; at least two of his teammates were also sidelined by MRSA.)

But the most striking differences between HA-MRSA and CA-MRSA involve the genes they carry. Antibiotic pressure in the hospital versus community settings selected for genes that encode different mechanisms of resistance. In this day of molecular typing of bacteria (a form of DNA fingerprinting), most CA-MRSA isolates circulating in the United States have been found to be related. Known as "clone USA 300," they carry the same package of genes that determine antibiotic resistance. The genetic makeup of different strains of HA-MRSA is much more diverse. (Interestingly, and somewhat unexpectedly, CA-MRSA isolates in Europe have a wide genetic diversity, unlike their American counterparts. That may eventually change, however, because a recent study demonstrated that the USA 300 CA-MRSA clone had made its way from the United States to Switzerland.)

There is one other very important difference between HA-MRSA and CA-MRSA: CA-MRSA can be killed by any of several oral antibiotics. HA-MRSA is resistant to all of these and can be defeated only by a very small group of antibiotics, most given intravenously.

Two new wrinkles in the MRSA epidemics have appeared in recent years. First, it is becoming more difficult to categorize cases as HA-MRSA versus CA-MRSA, as clones causing each epidemic have crept into each other's spaces. CA-MRSA is now invading hospitals, and HA-MRSA is being carried out into the community. Second, a new variant of MRSA has emerged in livestock (primarily pigs, but also cattle and poultry). This staph infection can be transmitted to humans as LA-MRSA (livestock-associated MRSA). But while many emerging infections like LA-MRSA are zoonotic (transmitted from animals to humans), a recent Norwegian study showed that pigs are at risk of acquiring MRSA from us (an example of a zooanthroponosis).[2]

The Enemy, Its Targets, and the Aftermath

I began my infectious diseases research career in 1975, studying mechanisms that *S. aureus* uses to evade what is the most effective antistaphylococcal "antibiotic" of all, namely, cells of the human immune system called neutrophils. Neutrophils (also referred to as polymorphonuclear leukocytes or PMNs) are members of the innate immune system that play a key role in the body's defense against *S. aureus*. You'll recall that you were introduced to neutrophils in chapter 4.

My research focused on one of the proteins in the cell wall of *S. aureus*, unimaginatively called protein A. This molecule interferes with the ability of neutrophils to recognize and engulf staph cells. But I soon realized that *S. aureus* is replete with what are called virulence factors. As you'll recall from chapter 4, virulence factors are features of microbes that turn them into our enemies. Virulence factors are what make us sick.

One common group of bacterial virulence factors is toxins. (You may have heard of one of them produced by a notorious *S. aureus* strain, staphylococcal toxic shock syndrome toxin-1, or TSST-1. Staphylococcal strains that release TSST-1 emerged in 1980 as the cause of toxic shock syndrome, which was associated with the use of high-absorbency tampons. This epidemic afflicted over two thousand women between 1980 and 1983.)[3]

After CA-MRSA emerged, researchers wondered why some patients develop extremely severe skin and soft-tissue infections, including a serious and swiftly spreading skin disease called necrotizing fasciitis, sometimes called flesh-eating bacteria by the media. More on this disease shortly.

Researchers also wondered why and how CA-MRSA could sometimes invade people's bloodstreams and spread to their lungs and other organs. We now know that clone USA 300 has a gene that encodes a toxin called Panton–Valentine leukocidin (PVL). But it remains unclear whether PVL explains the pathogen's virulence or whether it is just a marker associated with CA-MRSA.

Human skin provides a strong physical barrier to *S. aureus*. But once the physical barrier of skin epithelial cells is breached (by cuts, abrasions, and the like), *S. aureus* gains a potential opening. The staph may then provoke an inflammatory response—an influx of cells of the immune system, including neutrophils. This causes the disease cellulitis. (While *S. aureus* is the most notorious cause of cellulitis, other bacteria can also cause similar skin infections.)

The symptoms and signs of cellulitis include pain, tenderness, redness (known as erythema), swelling, and, sometimes, fever. Between a quarter and a half of all people with CA-MRSA come down with cellulitis. (Interestingly, many patients with CA-MRSA don't recall a break in their skin such as a cut or scrape. Often, though, they report what they thought was a spider bite, because the appearance of an initial skin lesion caused by CA-MRSA resembles a spider bite.)

Soft-tissue infection is a disease involving the tissues under the skin. This may involve an abscess or boil (painful collections of pus comprised of neutrophils). About half to three-quarters of patients with CA-MRSA have abscesses. Somewhere between 16 and 44 percent of people need to be hospitalized for these infections.

When a tissue under the skin called the fascia is invaded and dies (or, as healthcare professionals say, is necrotic), a life-threatening disease called necrotizing fasciitis develops. (A different microbe called Group A *Streptococcus* is another cause of this rare but very serious infection.) Even with proper treatment—the surgical removal of dead tissue, plus antibiotic therapy—40 percent of people who become ill with necrotizing fasciitis die. Physicians are trained to carefully watch for signs of this life-threatening infection—usually severe pain or delirium. Necrotizing fasciitis is a surgical emergency. Immediate incision and drainage, and the removal of all devitalized tissue, is critical. Antibiotic therapy is also necessary, but insufficient on its own.

Not surprisingly, skin and soft-tissue infections are the most common infections in hospitalized patients for which we infectious disease specialists are called in as consultants. We help the attending physicians decide what antibiotics should be used. But whenever necrotizing fasciitis is considered a possibility, what we recommend above all else is a scalpel.

Treatment and Prevention

As the number of cases of HA-MRSA cases mushroomed in the 1980s, vancomycin—an antibiotic developed about the same time as methicillin—became the mainstay of treatment. In the pre-MRSA era, methicillin and drugs like it were most commonly used because they had fewer side effects than vancomycin. But when HA-MRSA hit, only two antibiotics were left that worked: vancomycin and trimethoprim-sulfamethoxazole (TMP/SMX).

You may recall reading about vancomycin in chapter 13; it's a widely used agent for the treatment of *C. difficile* infection (CDI). For CDI, patients take vancomycin by mouth. But because vancomycin isn't absorbed through the gastrointestinal tract, it is of no value in treating systemic infections such as HA-MRSA or CA-MRSA. Thus, for MRSA infections, vancomycin must be given intravenously.

You can imagine the brief panic among physicians in 2002 when a vancomycin-resistant *S. aureus* (VRSA) strain was reported in a patient in Michigan. This strain also contained a gene for methicillin resistance, leaving a void of antibiotics to treat MRSA.

Fortunately, VRSA, and other strains that are only somewhat susceptible to vancomycin, are rare. We can also be thankful that the pharmaceutical industry stepped up to the plate in the 1990s by creating new antibiotics such as linezolid, quinupristin/dalfopristin, and daptomycin, which are all effective against MRSA, and in some cases also work against VRSA.

There are more choices for treating CA-MRSA, including vancomycin, TMP/SMX, clindamycin, and doxycycline. The last three of these can be taken orally. Because TMP/SMX is safe and relatively inexpensive, it is commonly used to treat CA-MRSA.

That said, antibiotics don't do much to treat abscesses. The reason why antibiotics don't work in these pus-filled sites that are loaded with neutrophils is unclear. So, for soft-tissue infections caused by CA-MRSA, surgical treatment—incision and drainage (I&D)—is often necessary. For uncomplicated boils, I&D by itself is usually sufficient. In other cases, I&D plus antibiotics is required.

Although developing a vaccine to prevent *S. aureus* infections has been a goal for some time, one doesn't yet exist. Thus, prevention of HA-MRSA in the United States depends on strict hand hygiene—proper hand washing and the wearing of gloves—and the isolation of patients who harbor the microbe.

Several European countries, including the Netherlands, Denmark, and Finland, have remained remarkably free of HA-MRSA. This appears to be due to their policy of isolating all patients when they are first admitted to the hospital and then testing them for MRSA. Patients are only allowed out of isolation if the test shows them to be MRSA free.

In the United States, one large hospital system in the Chicago area uses a similar approach to screening all incoming patients for MRSA. Patients found to harbor MRSA are placed in isolation; anyone who goes near them must wear a gown and gloves. This has dramatically reduced the rate of HA-MRSA infections. Given the high risk of surgical wound infections due to HA-MRSA, other hospitals routinely screen all patients before any surgery. If they are found to harbor MRSA, they are put in rooms isolated from other patients.

The good news is these strategies to defend hospitalized patients against MRSA are paying off. In 2013, a study by the CDC published in the *Journal of the American Medical Association*, reported a 54 percent decline in life-threatening HA-MRSA infections in the United States between 2005 and 2011, thus demonstrating the value of strict infection control measures. However, according to the CDC, the rate of decline in HA-MRSA infections slowed considerably between 2012 and 2017.[4]

Preventing CA-MRSA is an even more challenging problem. We can't eliminate, or even seriously reduce, activities that involve close human contact, or that can create breaks in the skin. This would mean, among other things, getting rid of most sports and many forms of exercise.

There are two things you can do to prevent CA-MRSA, however. First, if you do have a wound or break in your skin, wash it promptly and thoroughly with soap and water. (Recent studies suggest that the temperature of the water doesn't matter.) Second, if you see any evidence of cellulitis or soft-tissue infection, see a doctor as soon as possible.

Lessons for the Future

"No one is so brave that he is not disturbed by something unexpected."—Julius Caesar

Several decades ago, a world-renowned authority on antibiotics, who will go nameless, confidently made the claim at a conference I attended that *S. aureus* couldn't possibly become resistant to vancomycin. That prediction was proven wrong in 2002. For sure, that authority was chagrined. But the good news is that—as of late 2016—*S. aureus* hasn't found it easy to subvert vancomycin's action. Thus, it is still used extensively to treat MRSA infections. But we fear we are just sitting on a time bomb, and VRSA will eventually take off.

The further good news, however, is that the pharmaceutical industry has developed new drugs that work well against MRSA—for now. Several came to market in the 1990s; others have been added since. However, it seems likely that, given sufficient time, *S. aureus* will continue its track record of outsmarting us. (The next chapter deals with an even more alarming emergence of antibiotic resistance in other groups of microbes.)

The CDC ranks healthcare-associated infections as one of its top three public health concerns, right after alcohol-related harm and food safety. But here's a final bit of good news: as mentioned earlier, a recent study by the CDC published in the *Journal of the American Medical Association Internal Medicine* showed that invasive, life-threatening MRSA infections in healthcare settings are declining. Invasive MRSA infections declined 54 percent between 2005 and 2011, with 30,800 fewer severe MRSA infections.

In addition, the study showed nine thousand fewer deaths in hospital patients in 2011 versus 2005. While the reasons for these robust reductions aren't totally clear, it seems likely that improved infection control measures are at least partly responsible.

15

THE PERILS OF ANTIBIOTIC MISUSE

"I have always stuck up for Western medicine. You can chew all the celery you want, but without antibiotics, three quarters of us would not be here."—Hugh Laurie

"I do not wish to ban antibiotics or Cesarian sections any more than anyone would suggest banning automobiles. I ask only that they be used more wisely and that antidotes to their worst side effects be developed. The truth is always obvious in retrospect. How could people really have thought that the sun revolves around Earth or that Earth is flat? Yet dogma are powerful and to their adherents infallible."—Martin J. Blaser

ANTIBIOTIC RESISTANCE: THE NATURE OF THE CRISIS

In 2013, in her first annual report on combating the spread of antimicrobial resistance, Dame Sally Davies, Britain's chief medical officer, called for prompt global action. Davies said that, within two decades, antimicrobial resistance could cause tens of millions of patients to die following even minor surgery. Davies also said the problem is growing so large, and so serious, that the British government should rank it alongside terrorism and climate change as one of the country's biggest threats.

Those are strong words—and, as we will see in this chapter, they are well justified. Let's take a closer look—beginning with what the terms *antibiotic* and *antibiotic resistance* mean.

According to the *Merriam-Webster Dictionary*, the simple definition of an antibiotic is "a drug that is used to kill harmful bacteria and to cure infections." The full definition is "a substance produced by or a semisynthetic substance derived from a microorganism and able in dilute solution to inhibit or kill another microorganism."

The main cause of alarm is bacteria that fit the simple definition—the so-called superbugs. But as you read on, you will find that the phenomenon applies not just to bacteria but to the full range of pathogenic microbes—including viruses, protists, and fungi.

The term *antibiotic resistance* simply refers to the ability of microbes of any kind, via one of many genetically based strategies, to avert the effects of antibiotics.

Davies has not overstated the problem. Already large numbers of people die because of drug-resistant pathogens. A project supported by the nonprofit Wellcome Trust estimated in 2016 the annual worldwide death toll at seven hundred thousand. And without a new approach or new, more effective drugs, it was estimated that figure would rise to ten million by 2050. That's roughly one death every three seconds caused by antibiotic-resistant microbes—and almost twice the number of deaths that currently result from cancer.

In 2013, the Centers for Disease Control and Prevention estimated that infections caused by antibiotic-resistant bacteria resulted in more than two million illnesses and twenty-three thousand deaths in the United States annually. Currently, antibiotic resistance in America leads to more than eight million additional hospital days, an estimated $20 billion in added healthcare costs, and at least $35 billion in lost productivity each year. And an influential report on antibiotic resistance in 2016 estimated that the purely economic cost of lost global production from antibiotic-resistant bacteria could amount to a whopping $100 trillion by 2050, unless the problem is adequately addressed.

After analyzing data from 114 countries, the World Health Organization concluded in 2014 that antibiotic resistance poses a major global public health threat. They discovered antibiotic resistance "in every region of the world." The study focused on seven bacteria responsible for serious common diseases, including pneumonia, diarrhea, and bloodstream infections. The report suggests that we have entered a "post-antibiotic era" in which people are dying from simple infections that had been completely treatable for decades.

Few issues have captured our fear about the future more than the global problem of antibiotic resistance. Many infectious disease specialists, including myself, consider this the single biggest infectious disease threat that we currently face.

The magnitude of this problem is underscored by the WHO in its first surveillance report on antibiotic resistance in 2018. The WHO estimated that five hundred thousand people across twenty-two countries exhibited resistance to some categories of antibiotics. (At that time, only seventy-one countries had signed on to the Global Antimicrobial Surveillance System.)[1]

As in other modern-day pandemics, these antibiotic-resistant microbes are carried from country to country on the hands, or in the bowels, or in the genitourinary tracts, of unwitting travelers. And in some cases, animals or foods serve as vehicles for their spread.

The most frightening recent example of how bacteria superbugs get around is an *Escherichia coli* strain that is resistant to a "last-resort" antibiotic called colistin.[2] Colistin is used to treat infections caused by bacteria that the CDC considers "nightmare bacteria" because they are resistant to all other antibiotics—and kill about half the people they infect.

The first colistin-resistant *Escherichia coli* strain was identified by researchers in 2015, in a pig in China (where colistin is used on farm animals). In early 2016, the same researchers reported finding the gene that makes bacteria resistant to colistin, named *mcr-1*, in 15 percent of raw meat, 21 percent of animals, and 1 percent of hospitalized people tested in China. And in 2017 contaminated pet food was reported as a potential source of *mcr-1*-expressing bacteria in dogs and cats in Beijing.

Soon after the discovery of the *mcr-1* gene in China, it was detected in at least thirty other countries. In May 2016, the CDC announced the isolation of an *E. coli* strain carrying the *mcr-1* gene in a urine sample from a woman in Pennsylvania. In September, the CDC reported the fourth American infected with this highly resistant *E. coli* strain: a two-year-old girl in Connecticut who had traveled to the Caribbean and returned with it in her stools. And in January 2017, an *mcr-1*-carrying strain showed up in a patient at Los Angeles County Hospital. (It seemed most likely that this isolate was picked up by the patient in Asia.)

The really alarming aspect of the *mcr-1* gene is that it exists on what is called a plasmid—a small piece of DNA that is capable of moving from one bacterium to another, thereby spreading antibiotic resistance among different bacterial species. You'll recall reading in chapter 2 about this phenomenon, called horizontal gene transfer (HGT), as a driving mechanism for evolution. We'll return below to what it is about the *mcr-1* gene that makes it so scary.

How did this antibiotic-resistance crisis come about? To understand the root of the problem, let's return to some of the key insights from chapter 2.

Like all living creatures, microbes have been engaged in a struggle for existence ever since they first appeared on planet Earth some 3.8 billion years ago. To defend themselves against competitors, genes evolved that direct the production of antibiotics. The antibacterial drug penicillin, made by the fungus *Penicillium*, is one common example.

You'll also recall from chapter 2 that a teaspoon of soil contains about 240 million bacteria. Thus, it will come as no surprise that the search for new antibiotics has involved the screening of soil samples from around the world. (Over 80 percent of antibiotics in clinical use today originated from soil bacteria—either directly, as natural products, or as their semi-synthetic derivatives.)

In February 2018, a report in *Nature Microbiology* by investigators at Rockefeller University in New York City was heralded as a potential major breakthrough.[3] The researchers had discovered in soil samples a never-before-seen class of antibiotics they called malacidins (Latin for "killer of bad guys"). The reasons for the excitement: first, a truly new antibiotic hadn't been discovered since 1987; second, malacidins are effective against methicillin-resistant *Staphylococcus aureus* (MRSA); third, they appear to be nontoxic to humans; and, fourth, they were discovered using metagenomics. This opened a door for screening soil for other microbes that produce factors that weaken, damage, or kill other antibiotic-resistant bacteria.

But just as ordinary soil contains antibiotics, it's also replete with antibiotic-resistant genes. In one recent study, soil samples were found to harbor seven genes identical to those that enable bacterial pathogens to resist antibiotics. These same genes are behind microbes' antibiotic resistance to five major classes of drugs.

In another study, researchers from the University of Lyon in France analyzed bacterial DNA sequences from seventy-one different environments, including human feces, chicken guts, the ocean, and even Arctic snow. The researchers compared bacterial DNA extracted from these environments with sequences in the Antibiotic Resistance Database, which contains 2,999 gene snippets known to contribute to antibiotic resistance. Every environment they tested harbored an abundance of antibiotic-resistance genes.[4] In other words, antibiotic-resistant genes are everywhere.

Thus, at the same time that genes emerge for the production of new antibiotics, genes also evolve that provide mechanisms for resistance to antibiotics.

In some cases, like colistin-resistant *E. coli*, genes for resistance are carried on plasmids. In other cases, resistance genes reside in viruses that infect bacteria. In still others, they arise spontaneously in a bacterial chromosome.

The genetic alterations that confer resistance work in a number of ingenious ways. Some, such as the *mcr-1* gene, prevent the attachment of antibiotics to their target, like the binding of colistin to the cell wall of *E. coli*. Other gene products interfere with the action of antibiotics once they get inside a bacterium. Some actually pump antibiotics out of the bacterium. And then there are genes that encode for proteins (called enzymes) that zap antibiotics.

In short, *antibiotic resistance is a prime example of evolution in action*. The more antibiotics that we put into the environment, the greater the pressure on bacteria to develop resistance. Put another way, antibiotic pressure leads to the survival of the fittest. Thus, *the main reason for the antibiotic resistance crisis is a deluge of antibiotics in the environment*.

And guess who is responsible for this deluge? Sadly, two groups of *Homo sapiens*: physicians and farmers, both of whom dramatically overuse antibiotics.

As just one example, in America some forty million people are prescribed antibiotics annually for respiratory tract infections. Yet between half and two-thirds of them shouldn't receive antibiotics, because they likely suffer from viral infections, against which antibiotics are completely useless. Matters are far worse in most developing countries, where antibiotics are available over the counter without a prescription.

In both the developed and developing world, antibiotics are also widely used as growth supplements in livestock (mostly pigs, cows, and poultry). Worldwide, one hundred thousand tons of antibiotics are sold annually; more than half of these are used to fatten animals. In America, an estimated 70 to 80 percent of all antibiotics are used in agriculture, primarily to promote animal growth.

Not only do these antibiotics end up contaminating the environment, but recent studies show that when consuming food from livestock that were given antibiotics, you may ingest large numbers of antibiotic-resistant bacteria.

Another contributing factor to the antibiotic resistance crisis is a widening mismatch between supply and demand. During the first twenty-five years of my career as an infectious disease specialist (1975–2000), the pharmaceutical industry developed dozens of new antibiotics that proved effective against a wide variety of bacteria. But, beginning in 2000, just as the demand began accelerating because of antibiotic-resistant superbugs, the stream of new antibiotics slowed to a trickle.

The reasons why Big Pharma largely abandoned the $40-billion-a-year antibiotic market are complex. But one factor is profit. Today, the big money in drugs is for those taken for long periods of time, preferably for life, like statins and anti-hypertensives. Yet antibiotics are typically taken only for days or weeks. So pharmaceutical companies have focused their research largely on drugs that they can sell to the same people over and over. The consequence has been fewer new antibiotics to combat a superbug crisis.

WHAT ARE SUPERBUGS?

The term *superbug* was coined by the media to describe bacteria that are resistant to multiple antibiotics. You read in chapter 14 about one of the first such bacteria—MRSA. Like MRSA, all superbugs are highly lethal.

In earlier chapters of this book, you read about other bacteria that exacted a tremendous toll on humanity, such as *Yersinia pestis*, the cause of plague, and *Vibrio cholerae*, the cause of cholera. We also looked at emerging bacterial pathogens that kill a lot of people, such as *Legionella pneumophila*, the cause of Legionnaires' disease, and *Clostridoides difficile*, the main cause of death from diarrhea in the United States. But none of these bacteria are considered superbugs because they aren't resistant to multiple antibiotics.

Superbacteria

To this day, microbiologists lump the many different types of bacteria into two very large groups. These are differentiated according to their color when they are stained using a technique invented in the nineteenth century by a Danish bacteriologist, Hans Christian Gram. Under a microscope, stained gram-positive bacteria look violet, and stained gram-negative bacteria look red. (The basis for this is differences in the cell walls of gram-positive versus gram-negative bacteria.)

While antibiotic-resistant gram-negative bacteria have captured most of the limelight in recent years, two of the earliest bacteria superbugs that emerged toward the end of the twentieth century are gram-positive: MRSA and vancomycin-resistant enterococci (VRE), which is a common cause of urinary, intra-abdominal, and bloodstream infections. (A third gram-positive bacterium, *Streptococcus pneumoniae*—the number one cause of pneumonia—had everyone worried when its resistance to penicillin was reported. But so far, many other antibiotics have remained effective in killing it.)

Like healthcare-associated MSRA, VRE also emerged in hospitals. It is resistant to most penicillins and all cephalosporins, as well as many other antibiotics. But unlike MRSA, as its name implies, VRE is also resistant to vancomycin.

Fortunately, the pharmaceutical industry responded in the 1990s and early twenty-first century by releasing a number of FDA-approved antibiotics to treat these superbugs. These new antibiotics include linezolid, daptomycin, tigecycline, and several others.

In 1985, imipenem was approved. This penicillin-like antibiotic belongs to a class of drugs called carbapenems. Because they are active against almost all gram-positive and gram-negative bacteria, they were warmly welcomed by physicians. With some infections, carbapenems are our only effective weapon in the battle against antibiotic-resistant gram-negative bacteria. When these cease to be effective, we are in big trouble.

And sure enough, this trouble—emergence of carbapenem-resistant Enterobacteriaceae (CRE)—began to rear its ugly head in 2001 when the first *Klebsiella pneumoniae* strain that could destroy carbapenems was described in the United States. Later, this and other related CRE superbugs spread throughout the country, though some areas were hit harder than others.[5]

CRE are sometimes referred to as "nightmare bacteria" by health officials because they are not only resistant to carbapenems, but they also cause a wide range of life-threatening infections, including of the bloodstream, lungs, urinary tract, and abdomen, and infections following neurosurgery. These superbugs are increasingly common causes of these infections in hospitals and nursing homes. (Some of the most common CRE are strains of *E. coli, K. pneumoniae*, and *Enterobacter*.)

Currently, the real worry—verging on panic—about bacterial superbugs is the fear provoked by colistin-resistant *E. coli*. That is because colistin and a related drug, polymyxin B, are often the only antibiotics active against CRE.

So, when colistin-resistant *E. coli* emerged in 2015, alarms went off around the world. To make matters worse, in that same year another colistin-resistant gram-negative bacteria superbug, *Acinetobacter baumannii*, was reported as a cause of infection in twenty-two patients.[6] As mentioned earlier, resistance to colistin is conferred by a plasmid carrying the *mcr-1* gene, first described in China in 2015. And to increase our anxiety even further, in 2016 European scientists discovered in pigs a second colistin-resistance gene, *mcr-2*. The concern with *mcr-2* is that it may be passed among different bacterial species even more easily than *mcr-1*. Making matters even worse, in 2017 researchers in China and Europe discovered a third, *mcr-3*, and a fourth, *mcr-4*, mobile colistin-resistant gene in fecal samples from pigs.

The big worry, the one that is keeping public health leaders awake at night, is that one of these mobile genes will jump to CRE. While CRE isolates have been reported in livestock in Europe and Asia for some time, it first showed up in swine in the United States in 2016. Thus, the stage is set on farms for the emergence of this really frightening antibiotic-resistance scenario.

On top of all this, an *E. coli* isolate harboring the *mcr-1* gene was recently found in a patient hospitalized in New York City in 2015. And Italian researchers announced that they detected a *K. pneumoniae* carrying a variant of the *mcr-1* gene in a child with leukemia.

Other researchers added to the growing worries of zoonotic infection by identifying another potential animal-to-human connection. They recovered the *mcr-1* gene from the butts of seagulls in Lithuania and Argentina.[7]

Recently, strains of other gram-negative bacteria that cause typhoid fever, dysentery, and gonorrhea have emerged that are resistant to a wide range of antibiotics.

But the bacteria superbug that is the scariest of all—although rarely featured in our newspapers—is *Mycobacterium tuberculosis*, the cause of tuberculosis (TB). Of all the pathogens in the world, *M. tuberculosis* is the leading killer. Over the past two decades, two antibiotic-resistant strains have emerged—first, a multidrug-resistant TB (MDR-TB) strain, followed by an extensively drug-resistant TB (XDR-TB) strain.

Before an antibiotic to treat TB was introduced in the 1940s, this infection was a death sentence in half of all people sickened by TB. Even today, of the 10.4 million people who fall ill with TB every year, 1.8 million die. Because a large percentage of these patients live or die in the developing world, the ongoing devastation caused by TB is off most radar screens in wealthy countries.

When streptomycin, a highly effective treatment for TB, was introduced in 1944, it was heralded as a miracle drug. (The discoverer, Selman Waksman, received a Nobel Prize for Physiology or Medicine in 1952.) But very soon *M. tuberculosis* developed resistance to streptomycin. Fortunately, two new and highly effective antibiotics came along: isoniazid (INH) in the 1950s and rifampin in the 1960s. Together they paved the way for widespread and effective treatment of TB.

Over the course of several decades, additional anti-TB drugs were developed. But they were always added to regimens containing other antibiotics.

A fundamental principal in the prevention of antibiotic resistance was first learned in the treatment of TB: *use multiple antibiotics that hit different targets or act in different ways, at one time.* This makes it very challenging for the tubercle bacillus to develop resistance before it is killed.

Nonetheless, MDR-TB (strains that are resistant to at least INH and rifampin) and XDR-TB (strains that are resistant to INH *and* rifampin *and* two or more additional drugs) have emerged. Sadly, three out of four of the 480,000 cases of MDR-TB that occur around the world annually go untreated. It is estimated that MDR-TB could cost the world a shocking $16.7 trillion by 2050.

As for XDR-TB, it has now appeared in over ten countries. If this strain takes off, we may see a return to the pre-antibiotic era, when half of all TB patients died.

Fortunately, the havoc caused by MDR-TB and XDR-TB has been recognized by many governments. (In 2015, for example, President Obama launched a national action plan for addressing these two superbugs.) Nonprofits such as the Bill & Melinda Gates Foundation have also stepped in to attempt to address the issue. With their support, pharmaceutical companies are developing new anti-TB drugs that can more effectively kill and contain these mortal enemies. Results of clinical trials of newly developed three-drug regimens for XDR-TB are encouraging. And a recent report in the medical journal *Lancet* suggests that, with proper investments in diagnosis, treatment, and prevention, this ancient scourge could be conquered by 2045.[8] It is worth noting that there were other gram-negative bacteria superbugs well before CREs appeared. *Pseudomonas aeruginosa*, a gram-negative bacterium that is a common cause of hospital-associated infections, as well as a major threat to cystic fibrosis patients, is also resistant to many antibiotics. In the 1980s, a new group of bacterial enzymes called extended spectrum beta-lactamases (or ESBLs), were first detected. ESBL-producing gram-negative bacteria are resistant to penicillins and most cephalosporins. However, carbapenems are highly effective against these superbugs and have remained so since they were first introduced.

In 2017, the WHO for the first time released a list of drug-resistant bacteria that pose the greatest threat to human health. The bad news is that there are now twelve of them. (Three are gram-positive: MRSA, VRE, and *S. pneumoniae*—and the other nine are gram-negative: *Enterobacteriaceae, A. baumanii, P. aeruginosa, Shigella* spp, *Campylobacter* spp, *Salmonellae, Haemophilus influenza, Neisseria gonorrhoeae*, and *Helicobacter pylori*.)

Superviruses

Although viral infections are extremely common, antiviral drugs are available for a surprisingly small number of them. This is one reason why there are relatively few superviruses—that is, viruses that are resistant to multiple antiviral agents.

Unfortunately, the virus that sickens and kills the most people globally is one of these few superviruses: human immunodeficiency virus (HIV).

You read a brief history of anti-HIV treatment in chapter 7. You may recall from that chapter that the first antiretroviral drug, zidovudine, was introduced in 1987. Because HIV is an RNA virus that rapidly develops genetic mutations, resistance to zidovudine quickly emerged.

Taking a page out of the anti–*M. tuberculosis* treatment playbook, anti-HIV drugs that target other genes, and other mechanisms of viral growth in CD4 lymphocytes, were quickly developed. By 1996, highly active antiretroviral therapy (combinations of two or more anti-HIV agents) became standard. In 2019, at least twenty-six different highly effective anti-HIV drugs are on the market. On the downside, the emergence of highly drug-resistant HIV strains in the developing world, especially across Africa, Asia, and the Americas, has many people worried that we will eventually be in the same kind of fix with HIV that we are in with antibiotic-resistant bacteria.

Why has there been such tremendous success in dealing with this viral superbug, and so much less success in fighting bacterial superbugs? The answer, I believe, is largely money. Because none of the anti-HIV drugs eradicates the virus, HIV-infected people must be treated for life. And treatment is expensive: about $20,000 annually per patient in the United States, where more than a million people live with HIV.

Some resistance to drugs used to treat other viral infections—for example, influenza and infections caused by herpesviruses—has also emerged in recent years. But none of these viruses has achieved supervirus status—yet.

Superprotozoa

As you read in chapter 6, protozoa belonging to the genus *Plasmodium*—the cause of malaria—are among the all-time biggest killers of mankind. While considerable progress has been made in preventing and treating malaria in the twenty-first century, almost two hundred million cases of malaria still occur annually, and close to five hundred thousand people die from it each year.

The antimalaria drug quinine was introduced in the seventeenth century. Its use today is limited mainly by its toxicity rather than by any resistance developed by the malaria microbe.

Over the years, at least thirty-five drugs for the prevention or treatment of malaria have been introduced. However, some previously highly effective drugs, such as chloroquine, are rarely used today because the parasite has developed resistance to it in most parts of the world.

Artemisinin, a drug that is active against *Plasmodium falciparum*—the most lethal of the five species that cause malaria—was brought to market in 2004. Initially this drug was used alone. However, increasing resistance to it led to artemisinin-based combination therapy, which is now the norm. Thus, as with other superbugs, it's a game of catchup, with chemists constantly trying to keep up with microbial ingenuity.

Superfungi

In the latter part of the twentieth century, the number of patients with a compromised immune system (organ and bone marrow transplant recipients, cancer patients, and people with HIV/AIDS) skyrocketed. And opportunistic fungi came along for the ride.

To their credit, pharmaceutical companies stepped up to the plate and quickly developed a variety of new antifungal agents.

So far, the widespread use of antifungal drugs has spawned relatively few superfungi. But one such fungus recently debuted—*Candida auris*. This yeast, which was first encountered in the ear of a Japanese woman in 2009, has doctors very nervous. Since then it has cropped up throughout much of the world, usually in healthcare settings, where it causes bloodstream and wound infections.

In the United States, New York City has the most cases of infection caused by *C. auris* that are resistant to the commonly used antifungal drug fluconazole. An outbreak there in 2018, with a mortality rate of 45 percent, appears to be related to improper hospital infection control procedures. And in Oxford, England, a large outbreak of fluconazole-resistant *C. auris* infections in an intensive care setting was reported in the *New England Journal of Medicine* in October 2018.[9] The disease appears to have spread through the use of reusable temperature probes.

By mid-2017, at least 122 people in the United States had been infected by this fungus, most of whom died. *C. auris* is resistant not only to fluconazole but to two other classes of antifungal drugs as well. How this fungus could emerge, seemingly independently but simultaneously, in so many countries is a complete mystery—at least as of this writing in 2019.

THE NEW WAR ON SUPERBUGS

Despite recent reports of new antibiotics that kill pathogens without their developing resistance to it, experience suggests that searching for the Holy Grail—drugs that kill microbes in a way that makes it impossible for them to ever evolve resistance—is a futile effort. After all, microbes have had billions of years of experience in developing strategies that thwart antibiotics.

That said, there remain many mysteries. For example, it continues to baffle me why some bacteria like *Streptococcus pyogenes*, the cause of strep throat, remain uniformly susceptible to penicillin. Why hasn't it evolved resistance by now?

On a hopeful note, the gravity of the superbug crisis is now recognized by all the players. The rallying call to doctors, farmers, veterinarians, public health workers, government officials, pharmaceutical companies, investors, the food industry, and consumers is working. Even patients are beginning to get the message.

More good news: the emergence of superbugs is being viewed as a global problem. To raise and maintain awareness among all the stakeholders, the international nonprofit World Alliance Against Antibiotic Resistance was created in 2012. And for only the fourth time in its history, in September 2016 the United Nations General Assembly addressed a health crisis—antibiotic resistance.

Also encouraging is a growing awareness that antibiotic resistance should be approached from a One Health perspective. As you read in chapter 5, we are all in this together.

Because physicians are responsible for overprescribing antibiotics, measures are needed to curb this practice. Again, there is good news here. Antibiotic stewardship programs (ASPs) are now being implemented in most hospitals and are likely to become universal. Blue Cross Blue Shield recently reported that overall antibiotic prescribing is falling. This suggests that these programs are working.

ASPs are teams of knowledgeable physicians (often infectious disease specialists) and hospital pharmacists. They oversee antibiotic use in hospitalized patients, but they also have their eyes on outpatients, for whom most of the overprescribing takes place. Long-term care facilities, where overprescribing is also a big challenge, are beginning to follow suit.

The main goal of ASPs is to improve patient care. By advising what antibiotics are needed and *not* needed to treat infections, patient outcomes are improved. At the same time, drug toxicity, antibiotic resistance, and costs are all reduced. A 2016 report from the Pew Charitable Trust, *A Path to Better Antibiotic Stewardship in Inpatient Settings*, indicates that ASPs are working. Here are some other promising developments:

- Antibiotic stewardship is also being strengthened in veterinary medicine and animal agriculture. (Remember that about 70 to 80 percent of all antibiotics are used in agriculture, primarily to fatten up livestock).
- An increasing number of food companies are now creating, selling, and promoting antibiotic-free products. In 2017, Tyson, the largest chicken producer in the United States, and Sanderson Farms, the third-largest producer, announced they had stopped using antibiotics in their chickens. In September 2017, McDonald's, the biggest name in fast food, announced plans to cut antibiotic use in its worldwide chicken supply. (For an enlightening portrayal of how industrial chicken farms have misused antibiotics for decades—and have essentially bred antibiotic-resistant bacteria—read Maryn McKenna's excellent book *Big Chicken*.) And at about the same time, Burger King and KFC proudly reported that their food products were (and would continue to be) free of antibiotics. Consumer pressure is clearly behind these laudable changes.
- In 2015, under the leadership of the FDA, the U.S. government enacted the Veterinary Feed Directive Final Rule. Under this new rule, veterinarians provide a stewardship role for the use of antibiotics in animals. A major goal of this program is to phase out the use of antibiotics to promote animal growth. It should be noted that this strategy had already been enforced in several European countries.
- Efforts to curtail antibiotic resistance are also being coordinated by the WHO, the CDC, and many state departments of health in America, as well as by many public health organizations in other countries. An excellent review of the challenges that encourage the overuse of antibiotics and discourage new drug development are reviewed in the recent book *Superbugs: An Arms Race against Bacteria* by William Hall, Anthony McDonnell, and Jim O'Neill.

Adding to the challenge of antibiotic resistance, however, is recognition that in the developing world, lack of access to antibiotics is an even bigger problem. Currently in these countries, limited access to antibiotics causes more deaths than antibiotic resistance. For example, in 2018 the results of a large placebo-controlled study of azithromycin in sub-Saharan Africa, published in the *New England Journal of Medicine*, showed that childhood mortality was significantly lower in communities randomly assigned to mass distribution of this antibiotic.[10] It appeared that the basis for this benefit was related to a beneficial alteration of the gut microbiome in people who received azithromycin.

It is clear that without government support and the political will to reduce or halt antibiotic resistance, little would be accomplished. In the United States, a March 2015 executive order by President Obama doubled the funding to fight antibiotic resistance. Later that year, a panel of experts, the Presidential Advisory Council on Combating Antibiotic-Resistant Bacteria, held its first meeting. But sadly, as I write this paragraph in early 2019, the fate of these important initiatives is uncertain.

In 2016, the United Kingdom Review on Antimicrobial Resistance called for $40 billion to tackle antibiotic resistance. And both the G7 and G20 groups, together with the WHO, called for big changes in how antibiotic research and development are financed.

Most promising of all, in July 2016 the creation of an international partnership, the Combating Antibiotic-Resistant Bacteria Biopharmaceutical Accelerator (CARB-X) was announced. CARB-X supporters come from multiple government agencies, health charities, and private partners in the United States and Britain. They pledged several hundred million dollars of funding over the next five years to accelerate the process of bringing promising new antibiotic candidates to healthcare providers.

The most important group of stakeholders in the fight against the overuse of antibiotics—the general public—has been the slowest to get on board. While practitioners are the ones who inappropriately write prescriptions for antibiotics for viral infections (sometimes due to real or perceived pressure from patients who want these "miracle drugs"), we would like to see this habit reversed by patients who ask for the evidence they have a bacterial infection when an antibiotic is prescribed. What is needed now is for a "Stop Overuse of Antibiotics" campaign that includes patients and nonpatients alike.

Fortunately, the CDC and other public health organizations provide valuable information to inform the nonmedical community. For example, take a look at the "Get Smart About Antibiotics" link on the CDC's website.

What can you do personally to protect yourself from superbugs? Superbugs, like those that aren't so super, are largely carried on human hands. So as your first and most important step: *thoroughly wash your hands with soap* after visiting the bathroom, after changing a diaper, before preparing foods, and before eating. (For more detailed and exact instructions on hand washing, Google the WHO's six-step hand hygiene technique, or the CDC's slightly simpler three-step technique.)

Second, remember that antibiotics are of no value in the treatment of viral infections—for example, viral pharyngitis (the cause of most sore throats), acute sinusitis, acute bronchitis, and acute otitis (middle ear infection). Using antibiotics will not just be useless; it will promote bacterial resistance and may cause serious side effects, such as allergies and *C. difficile* colitis, which kills about fifteen thousand Americans every year (see chapter 13).

We now live in an era of patient-centered care. As a patient, you need to be in the driver's seat on all decisions regarding your treatment. *So if your doctor tells you that you don't need an antibiotic and explains why, thank them!* And if you're asked to rate their level of care, give them credit for it.

If and when you *are* prescribed an antibiotic, ask not only why. Also ask if an equally effective nonantibiotic treatment is available. If you're not sure whether a particular drug is an antibiotic, just ask.

And if you find yourself in the hospital, always, always, *always* ask everyone who enters your room to wash their hands as soon as they arrive—even before they hug you hello.

Part Three

Germs in the Future

16

THE STRAIGHT POOP ON FECAL TRANSPLANTS

"You aren't what you eat—you are what you don't poop."—Wavy Gravy

FECAL MICROBIOTA TRANSPLANTATION (FMT)

"Poop is one of the most wholesome, beautiful and natural experiences that money can buy."—Steve Martin

My most memorable time in medical school was when, as a fourth-year student, I took a clinical elective at a hospital in a small town in South Korea. It was 1970, and South Korea was then a developing country. My wife accompanied me, and while we were there, she carried out a surveillance study of stools from hospitalized patients. Using a microscope, she was looking for the eggs of parasites. She found those eggs in almost every stool sample she examined. This implied that many people in South Korea were infected with some kind of parasite.

The explanation for this was close at hand. Every morning, we watched as a worker made his rounds carrying a "honey bucket" to collect "night soil." Those were the euphemisms we Americans used. He was collecting human feces from privies, to be used as fertilizer in nearby fields.

My wife and I had the good fortune to return to South Korea in 2005. By then, South Korea had become one of the most developed countries in the world. The collection of night soil, and the widespread use of outdoor privies, were long gone. And so were the parasites.

In 1970, the furthest thing from my mind was that one day human feces would be used to treat sick people. Nor was I aware then of how manure (whether from animals or humans) encourages soil microbial activity for the benefit of plants—and, indirectly, the creatures that eat them, including *Homo sapiens*.

What Is FMT and How Does It Work?

Fecal microbiota transfer, also known as fecal bacteriotherapy, stool transplantation, and FMT, is the process of transplanting feces from a healthy donor into a recipient. This may sound crazy, primitive, or barbaric, but it's none of the above. The fact is that it works.

The idea of treating gastrointestinal ailments with feces dates back to fourth-century China. Twelve centuries after that, Li Shizhen, a famous Chinese physician, used "yellow soup" containing fecal matter to treat severe diarrhea. And during World War II, German soldiers confirmed the benefit of a Bedouin remedy for dysentery—consumption of fresh camel feces.

The first description of FMT, however, was published in 1958 by a Colorado surgeon, Ben Eiseman, and his colleagues. They had successfully treated four critically ill patients who suffered from serious psuedomembranous colitis.

It would take another two decades before the cause of psuedomembranous colitis was established—the bacterial pathogen *Clostridioides difficile*, which you read about in chapter 13. (You'll recall that its genus name recently was changed from *Clostridium*.) *C. difficile* emerges in the colon following the administration of antibiotics that knock out competing, friendly bacteria.

To understand how FMT works, let's return to some of the insights from chapter 3, which focuses on the human microbiome. If you're healthy, your gut microbiome is made up of about thirty-nine trillion bacterial cells from about two thousand bacterial species. These bacteria typically belong to four phyla: Firmicutes, Bacteroidetes, Actinobacteria, and Proteobacteria. It also contains enormous numbers of archaea, viruses, fungi, and protists—but our current thinking is that FMT is all about bacterial warfare: pitting good bacteria against mortal enemies such as *C. difficile.*

FMT works by helping a human gut with dysbiosis (a microbial imbalance) restore a healthy balance of microbes (or eubiosis). Scientists believe it does this by colonizing the gut with friendly bacteria to outcompete any malicious bacteria. Given the extraordinary complexity of the gut microbiome, it may eventually turn out that this theory is overly simplistic. Nonetheless, over the past seven years, research has established that FMT is a highly successful therapy for recurrent *C. difficile* infection (CDI).

The potential application of FMT goes far beyond treatment of gastrointestinal infections. As mentioned in chapter 3, many human maladies, including type 2 diabetes, obesity, cardiovascular disease, Crohn's disease, ulcerative colitis, irritable bowel syndrome, some cancers, certain autoimmune diseases, asthma, allergies, and even certain neuropsychological disorders, may be rooted in dysbiosis.

FMT 101

These are still relatively early days for FMT research. Nonetheless, the evidence that this is a bona fide treatment of recurrent CDI has accumulated at lightning speed. The first randomized controlled trial of FMT for CDI was carried out in the Netherlands. The results, published in 2013 in the *New England Journal of Medicine*, demonstrated that FMT was superior to the antibiotic vancomycin given to the control group.[1] This study supported the results of a number of earlier tests and anecdotal reports.

You can imagine the hurdles researchers need to clear to carry out these studies—and to get patients to agree to FMT treatment. One of the first hurdles, is the so-called ick factor or snicker factor. But researchers showed as early as 2012 that patients are open to considering FMT as a treatment for recurrent CDI, especially when it is recommended by a physician.

Remember that an estimated five hundred thousand people in the United States develop CDI each year, and about fifteen thousand of them—roughly 3 percent—die. While antibiotic therapy works as an initial treatment in most cases, the disease recurs in at least 25 percent of patients. All of these patients are miserable with diarrhea, abdominal pain, and systemic symptoms. While they are usually treated with another course (or two or three) of antibiotics, when they are offered FMT as a potentially more effective treatment, many jump at the opportunity.

A second and much more challenging series of hurdles relates to regulatory issues. In the United States, this means getting approval from the Food and Drug Administration—first to carry out clinical trials, and then, if everything goes well, to be permitted to routinely use the procedure as a form of treatment. (The FDA classifies human stool as a biological agent. To ensure patient safety, the FDA regulates its use in FMT therapy and other research.) Currently, U.S. doctors using FMT to treat recurrent CDI aren't required to obtain an investigational new drug (IND) permit—but the FDA strongly encourages them to do so.

Here are just some of the matters that researchers need to consider:

Issue 1: Where does the donor feces come from? Currently, at most centers where FMT is performed, donor stools are provided by healthy people, often sympathetic family members. Donors are screened to make sure they don't have diseases associated with an unhealthy microbiome. Their feces is also routinely tested for potential pathogens. (Fortunately, frozen donor stool has been shown to work just as well as fresh. Because of the practical difficulties of obtaining and processing stools, a nonprofit stool bank, OpenBiome, was developed at the Massachusetts Institute of Technology. As far as we can tell, this stool bank provides safe fecal matter for FMT treatment.)

Issue 2: What is the route of administration? So far, FMT has usually involved infusion of fecal material into the patient's body in one of two ways: by tubes through the nose into the small intestine, or into the colon via colonoscopy or enemas. No surgery is required. (Oral administration in capsules—so-called poop in a pill—would naturally be preferred by both patients and healthcare professionals, but this form of FMT is still being studied and hasn't yet been approved. However, results of studies on oral treatment, carried out by researchers at the University of Calgary and elsewhere, are encouraging.)

Issue 3: What is the biological nature of the material administered? As we learn more about the human microbiome, more sophisticated treatments are emerging. For example, one variation being studied is the administration of bacterial spores rather than actively growing bacteria. As mentioned in chapter 13, these studies showed that spores seem to be effective in treating recurrent CDI. (Patients were given spores from a

C. difficile strain that doesn't produce the toxins that cause CDI. This resulted in a significant reduction of recurrences of their CDI. It appears that this non-toxin-producing relative of toxic *C. difficile* takes over its space in the colon.)

Equally promising are the results of another FDA-approved clinical trial that appeared in 2016 in the *Journal of Infectious Diseases*.[2] For this trial, spores from about fifty species of the bacterial phylum Firmicutes were put into capsules, which patients swallowed. Almost 90 percent of the patients treated with these spores got better.

FMT is not a magic bullet or miracle procedure, however. Some failures have been reported. In one study, FMT worked about 75 percent of the time—certainly far better than no treatment, but also far worse than what healthcare professionals want to see. (Failure was most likely to occur among hospitalized patients. This suggests that the oldest and sickest people with CDI are the ones for whom FMT may not work.) A 2017 review by Stuart Johnson and Dale Gerding, infectious diseases experts in this field, of six randomized controlled trials led them to conclude that refinement of FMT is clearly needed "to make it a more acceptable, safe, and more defined product."[3] More disappointing, however, are the results of a recent clinical trial reported in the *Atlantic*, in an article titled "Sham Poo Washes Out."[4] Supported by a young venture company, Seres Therapeutics, eighty-nine patients with recurrent CDI were enrolled in this study of their lead product—SER-109—consisting of a single capsule containing one hundred million spores from fifty species of gut bacteria, or a placebo. Sadly, the early hopes pooped out in July 2016 when the results showed no benefit versus the placebo ("sham") pill. The explanation for this surprising failure is currently unclear. (The fact that the company embarked on a second clinical trial of SER-109 in 320 patients, registered in June 2017 in ClinicalTrials.gov, suggests to me that they found a reason for the initial failure, which they plan to circumvent this time around.)

Also, while FMT is generally considered safe, complications occasionally do occur. The recent death of an FMT recipient caused by a multidrug-resistant *Escherichia coli* acquired via the transplant prompted the FDA to issue a warning that donor stool be tested for multidrug-resistant bacteria.

Of particular interest is a report in 2015 of a thirty-two-year-old woman with recurrent CDI who underwent FMT. At her request, her sixteen-year-old daughter, who weighed 140 pounds and had a body mass index (BMI) of 26.4 but was otherwise healthy, served as the stool donor. At the time of the FMT, the patient weighed 136 pounds and had a BMI of 26. (A BMI of 18.5–24.9 is considered normal, so both donor and recipient were deemed slightly overweight.) The good news was that the FMT cured the recipient. The bad news was that, over the next sixteen months, she gained thirty-four pounds—while eating a medically supervised diet and following an exercise program. In the twenty months after that, she gained another seven pounds. Was her weight gain somehow caused by (or related to) the fecal transplant? We don't know for certain, but it is likely. You'll recall from chapter 3 the potential link between the gut microbiome and obesity.

FUTURE APPLICATIONS OF FMT

> "I was aware that an entire new science was being born. And actually I was almost salivating with envy—Boy, I wish I was in that field! And it just so happened that as a gastroenterologist, I'm in the middle of that field. So I couldn't resist entering it. We're at the beginning of this new science. This is a wide-open new frontier."—Alexander Khoruts

Alex Khoruts, medical director of the University of Minnesota Microbiota Therapeutics Program, is one of the pioneers of FMT. He and two renowned scientists, environmental microbiologist Michael Sadowsky and computational microbiologist Daniel Knight, have worked together to help us understand how FMT works and to provide insights into how it might be used. They are my colleagues, so I've had the pleasure of attending their lectures. While their enthusiasm for the fields of FMT and microbiome research is palpable, they are quick to caution that the science and clinical application of FMT are still in their infancy. And all three underscore that correlation doesn't mean causation.

The swift, successful development of FMT into a proven therapy for recurrent CDI has fueled excitement for its potential in treating other ailments associated with a disturbed gut microbiome. But it could be that starting with CDI was fortuitous. It is a disease caused by a known mortal enemy—*C. difficile*. And the evidence so far suggests that other bacteria in our gut naturally work to eliminate it.

But there are innumerable other illnesses that may be caused by dysbiosis—and for the vast majority of them, we don't know any of the relevant microbial enemies *or* friends. Furthermore, the human gut is colonized by an incredibly large array of bacterial species, as well as by archaea, viruses, fungi, and protists. We could be searching for a handful of needles in an almost unimaginably immense haystack.

That said, it is nonetheless inspiring to see the large number of clinical trials of FMT registered with ClinicalTrials.gov, a service of the U.S. National Institutes of Health. Many of these trials have provided evidence to support the more widespread value and application of FMT.

The controlled clinical trials underway target conditions that include obesity, metabolic disorders, ulcerative colitis, Crohn's disease, irritable bowel syndrome, Parkinson's disease, and autism.[5] Given the associations established between the microbiome and cancer immunotherapy (see chapter 3), it is not surprising that FMT is also making its way into the field of oncology.[6]

Should sufferers with recurrent CDI, or patients with disorders associated with dysbiosis, consider FMT? My advice: consult your doctor. But if I had recurrent CDI, I would ask for a referral to a specialist with an established track record of using FMT.

For all other patients with known or suspected dysbiosis, my advice is stay tuned. The field of FMT is currently something like the Wild West in the mid-1800s. Online, you can find do-it-yourself fecal transplants, with accompanying YouTube videos. Please don't do it yourself. (Even in the best of hands, FMT doesn't always work and isn't without complications.) You can also connect with cultlike groups that promote FMT for many unproven purposes. Please avoid these as well.

Genuine answers to questions about the value of FMT will come only from properly run randomized clinical trials. These are being conducted as I write this chapter. So keep an eye out for new poop on poop.

17

HEALING WITH FRIENDLY BACTERIA AND FUNGI

"And let me adde, that he that throughly understands the nature of Ferments and Fermentations, shall probably be much better able than he that Ignores them, to give a fair account of divers Phaenomena of severall diseases (as well as Feavers and others) which will perhaps be never throughly understood, without an insight into the doctrine of Fermentation."— Robert Boyle, eighteenth-century chemist

WHAT ARE PROBIOTICS AND HOW DO THEY WORK?

You'll recall reading in the "Intimate Friends" part of this book that the microbial community in your gastrointestinal tract is made up of about forty trillion bacterial cells, which belong to about two thousand different species. Additional inhabitants include members of more than one hundred fungal species, mainly different kinds of yeast. And, assuming you're healthy, virtually all these germs are either commensals (benefiting themselves but not giving you any trouble) or mutualists (benefiting both themselves and you).

Probiotics are all about the mutualists—our intimate friends.

The word *probiotic* comes from the Greek words *pro*, meaning "promoting," and *biotic*, meaning "life." Probiotics are defined by the Food and Agricultural Organization of the United Nations and the World Health Organization as "live microorganisms which, when administered in adequate amounts, confer a health benefit on the host."

In addition to a wide variety of probiotics, it's also possible to buy prebiotics—nondigestible carbohydrates that nourish our intestinal microbiota—and synbiotics—products that contain both probiotics *and* prebiotics.

Probiotics are naturally found in a variety of foods (most commonly yogurt and kefir) and can also be purchased as dietary supplements. Although none has yet been approved by the U.S. Food and Drug Administration, probiotics, prebiotics, and synbiotics are widely used for their perceived health benefits.

I say "perceived" instead of "proven" because the FDA has so far withheld its blessing—but there is plenty of evidence to suggest that these friendly bacteria and fungi do help our bodies. Microbial cultures were used for thousands of years to ferment food and alcohol—many centuries before Louis Pasteur discovered germs in the nineteenth century.

In fact, before turning his attention to the germs that cause diseases, Pasteur determined in the 1850s and 1860s that some of them are responsible for the process of fermentation. Pasteur also discovered that heating beer and wine was enough to kill most bacteria that caused spoilage. This heating process, which became known as pasteurization, continues to be widely used today to kill potentially harmful microbes in milk, bottled juices, and other foods.

Most people know that beer, wine, yogurt, cheese, sauerkraut, and leavened bread are all fermented. But they may not know that germs are behind the fermentation. In these foods, certain types of yeast and bacteria convert sugar to acids, gases, or alcohol.

Elie Metchnikoff—the founder of cellular immunology, whom Pasteur recruited to the Pasteur Institute, and whom you met in chapter 4—is widely regarded as the father of probiotics. Metchnikoff suggested that harmful bacteria inhabiting our lower intestine release toxins that are responsible for senility. Later, he championed the broader notion that a microbial imbalance (a deficiency of good germs, or an overgrowth of harmful ones) in the lower intestine was an underlying cause of much ill health. Strange as this sounds, increasing evidence suggests that Metchnikoff was correct.

In the late 1800s, when Metchnikoff was doing his research, Bulgarians were known for their longevity. (Today, Bulgarians' life expectancy is only average.) At the time, Bulgarians were also well known for their high consumption of yogurt. Metchnikoff proposed that eating the lactic-acid-producing bacteria in fermented dairy products such as yogurt supported the health and longevity of Bulgarian peasants. He suggested replacing harmful microbes in our guts with friendly bacteria—in a process called orthobiosis—in his 1907 treatise *The Nature of Man: Studies in Optimistic Philosophy*. He found that yogurt, particularly the Bulgarian variety, contained two types of bacteria that killed harmful bacteria, and this appeared to underlie its medicinal properties.

In the early years of the twentieth century, this idea was twisted and taken to harmful extremes by surgeons who removed parts of people's intestines. Another unproven, though much gentler, approach was intestinal cleansing, also called colon hydrotherapy. So-called colonics are recommended to this day by some holistic healers, despite the absence of evidence that they are beneficial and safe. Neither of these drastic approaches to ridding the gut of harmful bacteria turned out to help. But Metchnikoff's basic idea that what lives in our gut has a profound effect on our health remains medically and scientifically sound.

There are three ways probiotics appear to work in the gut: by outcompeting or knocking out pathogens, by enhancing the integrity of the gut lining, and by suppressing inflammation.

To be effective, the microbes in probiotics must survive the stomach's acid environment and bile salts in the upper intestine. Even then, they don't stay alive and anchored in the gut for long, so we need to continuously consume most probiotics.

Kefir is claimed to be an exception. The microbes in kefir are said to stay put, apparently colonizing the intestinal tract. But scientific evidence backing this claim is difficult to find.

Most commercially available probiotics contain at least eight different species and strains of the most commonly used bacterial genus, *Lactobacillus*. Other common probiotic bacteria include *Bifidobacterium*, *Streptococcus*, and *Escherichia*. The most common beneficial fungus in probiotics is *Saccharomyces boulardi*.

Most probiotics are aimed at correcting problems in the gut, but other organs may indirectly receive benefit, such as the brain, vagina, respiratory tract, and skin.

ARE PROBIOTICS EFFECTIVE?

"In theory, there is no difference between theory and practice. In practice there is."—Yogi Berra

The world market for probiotics in 2016 was estimated at about $46 billion. Yet none of the hundreds (and perhaps thousands) of probiotics on the market has gained FDA approval, either as a treatment or as a preventative for any health problem. (All probiotics are recognized by the FDA as dietary supplements.) Underscoring their popularity, a recent nationwide survey of 145 American hospitals found that patients were prescribed probiotics in 96 percent of them. And in a health survey in 2012, 3.9 million American adults reported using probiotics or prebiotics. Since then, that number appears to have dramatically increased.

So why, given all the enthusiasm for the health benefits of probiotics, don't we have any that are FDA approved? After all, the theoretical basis for their effectiveness—including emerging scientific support from the Human Microbiome Project—has been clearly established. I suspect that the answer is money.

A recent study from the Tufts Center for the Study of Drug Development estimates that the cost of developing a prescription drug that receives FDA market approval is $1.4 *billion*. A good share of this cost relates to the FDA's regulatory requirements during testing. Because development costs are so high, many potential cures and preventive measures get abandoned or ignored rather than fully tested.

Despite these practical challenges, a surprisingly large number of studies of probiotics have been carried out. In the early 1990s, what is called evidence-based medicine—the backbone of which is randomized clinical trials (tests conducted under rigorous conditions), or RCTs—became the gold standard for assessing the risks and benefits of drugs and dietary supplements.

Although there is clearly a need for further RCTs, preliminary studies suggest that probiotics may help treat—and prevent—certain forms of diarrhea, such as *Clostridioides difficile* infection, the very serious gastrointestinal tract infection you read about in chapter 13.

Probiotics also appear to be effective in treating gastroenteritis due to rotavirus, a common cause of severe diarrhea worldwide. Necrotizing enterocolitis, a life-threatening disease that affects the bowel in premature infants, also appears to respond to probiotic therapy. (But the negative results of a recent large RCT of the probiotic *Bifidobacterium* BBG-001 in very preterm infants suggest that not all probiotic strains prevent this serious condition.)[1]

The negative findings of large placebo-controlled RCTs of *Lactobacillus*-based probiotics for the treatment of acute gastroenteritis in children, reported in the *New England Journal of Medicine* in November 2018, were also disappointing.[2] And the results of recent studies by researchers in Israel have raised questions about the widespread use of probiotics to impart wellness and to restore the gut microbiome after antibiotic use.[3]

Without doubt, the most promising study showing the benefit of probiotics in preventing life-threatening infections in newborns was reported in the journal *Nature* in 2017.[4] In this RCT, carried out by Pinaki Panigrahi and his colleagues in rural India, a synbiotic (a combination of *Lactobacillus plantarum* and a prebiotic known as fructooligosaccharide) was given to newborns. After sixty days of monitoring, they found that this regimen significantly reduced the number of bloodstream infections and the number of deaths. If the results of Panigrahi's study are replicated in other developing countries, this would be a major medical breakthrough.

In 2017, one in five Americans took probiotics for digestive problems. Irritable bowel syndrome (IBS) is one such bowel disorder, with symptoms that commonly include pain or discomfort; bouts of diarrhea, constipation, or both; and bloating or bowel distention. In the United States, IBS afflicts between 3 and 20 percent of adults. While the cause is unknown, an altered gut microbiome is thought to play a role. So far, studies of probiotics as a treatment for IBS are encouraging, showing significant reduction in abdominal pain and other symptoms.

Probiotics may improve human health in other ways. In multiple studies, probiotics achieved better results than placebos in preventing upper respiratory tract infections, such as the common cold. Other recent studies of probiotics, including yogurt, suggest that they may slightly lower blood pressure in patients with hypertension. Preliminary results also suggest that probiotics can help treat or prevent many other medical conditions, including oral and vaginal infections caused by the yeast *Candida*, breast infections associated with breast feeding, bacterial vaginosis, hepatic encephalopathy, hypercholesterolemia, allergies, and eczema. But, as of this writing in June 2019, we don't (yet) have clear evidence of this from well-designed randomized controlled studies.

For an excellent review of how probiotics may improve your mood, I recommend the book *The Psychobiotic Revolution: Mood, Food, and the New Science of the Gut-Brain Connection*.[5] The book is coauthored by an American science writer, Scott Anderson, and two researchers at University College in Cork, Ireland—John Cryan, a neuroscientist, and Ted Dinan, a psychiatrist. They explain how gut bacteria converse with the brain and the implications of recent research findings for the treatment of mental disorders such as depression and anxiety.

Underscoring the difficulty in understanding how probiotics might work, however, a recent review by University of Copenhagen researchers of seven RCTs found that probiotics *didn't* change the fecal microbiota of the healthy subjects who took them.[6] This is not surprising, since the bacteria in probiotics typically don't colonize the gut.

ARE PROBIOTICS SAFE?

Although probiotics are generally regarded as safe, researchers at the Celiac Disease Center at Columbia University Medical Center recently reported that 55 percent of the high-selling probiotics they tested contained gluten. So gluten-intolerant folks—especially patients with celiac disease—need to be careful about the probiotics they ingest.

In 2015, the results of a large study by the federal government on the dangers of dietary supplements were published in the *New England Journal of Medicine*.[7] The findings suggest that injuries caused by dietary supplements lead to more than twenty thousand emergency room visits and two thousand hospitalizations per year. And in November 2015, the Justice Department filed criminal and civil enforcement actions against 117 companies and individuals involved in selling of workout supplements that contained an amphetamine-like stimulant. Equally disturbing is a study published in 2018 that reveals the underreporting of harmful effects of probiotics, prebiotics, and synbiotics, even in randomized controlled studies.[8]

These reports underscore the importance of monitoring dietary supplements for dangerous side effects. They also remind us of the importance of carefully reading labels on all dietary supplements. But reading the labels may not be good enough as ingredients or contaminants that aren't listed sometimes turn out to be the cause of ill effects.

As we've seen, the process of gaining FDA approval for a probiotic food product or supplement is both complicated and expensive. For most physicians, including me, however, this approval is an essential step before we are likely to recommend a particular probiotic's use. Nonetheless, the fact that probiotics are widely used, even within American hospitals, suggests that wishful thinking sometimes trumps a lack of clear scientific evidence.

18

HEALING WITH FRIENDLY VIRUSES

"We live in a dancing matrix of viruses; they dart, rather like bees, from organism to organism, from plant to insect to mammal to me and back again, and into the sea, tugging along pieces of this genome, strings of genes from that, transplanting grafts of DNA, passing around heredity as though at a great party."—Lewis Thomas, American physician and writer

GETTING TO KNOW PHAGES

"On opening the incubator I experienced one of those rare moments of intense emotion which reward the research worker for all his pains: at first glance I saw that the broth culture, which the night before had been very turbid was perfectly clear: all the bacteria had vanished . . . as for my agar spread it was devoid of all growth and what caused my emotion was that in a flash I understood: what causes my spots was in fact an invisible microbe, a filterable virus, but a virus parasitic on bacteria. Another thought came to me also. If this is true, the same thing will have probably occurred in the sick man. In his intestine, as in my test-tube, the dysentery bacilli will have dissolved away under the action of their parasite. He should now be cured."—Félix d'Herelle

The word *bacteriophage*—from *bacterio* (bacteria) and the ancient Greek word *phagein* (meaning "to eat")—was coined by one of the tiny creatures' discoverers, the French microbiologist Félix d'Herelle.

Bacteriophages, often referred to as phages, are an enormous group of viruses that infect bacteria. As we saw in earlier chapters of this book, bacteriophages are found in all environments where bacteria or archaea exist. This means—as Carl Zimmer points out in his book *A Planet of Viruses*—they live just about everywhere on our planet. [1]

As you'll recall, scientists estimate that bacteria outnumber and outweigh all animal life on Earth. Phage biologists calculate there are even more viruses on our planet, and most of these are phages. This makes viruses by far the most numerous creatures on Earth. And phages, in particular, are hardworking: biologists estimate that they destroy about half the bacteria in the world *every forty-eight hours*.

Bacteriophages were first postulated in 1896, when Ernest Hankin reported that something in the waters of the Ganges River in India seemed to harm the bacterium that causes cholera, *Yersinia cholerae*. Whatever it was, Hankin knew it was very, very small because it readily passed through a fine porcelain filter that trapped all bacteria.

The credit for actually discovering and identifying phages is shared by the British bacteriologist Frederick Twort (1915) and Félix d'Herelle, both of whom did pioneering work in the 1910s. D'Herelle also came up with the concept of phage therapy—using phages to cure or prevent disease.

Two kinds of phages infect bacterial cells: lytic phages and lysogenic phages.

Lytic phages break open bacterial cells, multiply inside them, and then immediately destroy them. Then the phage progeny move on to find new bacterial cells to infect.

In contrast, the genomes of lysogenic phages become integrated into the DNA of the host bacterial cell. There they replicate harmlessly for a time—or else become established inside the cell as a separate DNA molecule known as a plasmid. In either case, the phages do no immediate harm. In fact, sometimes lysogenic phages actually benefit their host by adding new functions to its genome. But when the conditions of the host cell deteriorate—often because of damage to it—the phages take advantage of its vulnerability by becoming active, replicating, and breaking it open, killing it.

Any one type of phage infects only very specific bacteria. This phenomenal precision makes phages highly attractive as therapeutic agents, because they take down certain bacterial enemies while leaving our bacterial friends intact. In contrast, antibiotics inhibit or kill trillions of bacteria of all types—our enemies, our friends, and lots of harmless, innocent bystanders.

PHAGE THERAPY

"Enough of all this mere frittering and vanity. Let's really cure somebody!"—Dr. A. Dewitt Tubbs, in Sinclair Lewis's novel *Arrowsmith*

In the 1920s and 1930s, physicians used phages to treat a variety of infectious diseases. They were regularly sold by pharmaceutical companies such as Eli Lilly & Company, and their popularity was highlighted in Sinclair Lewis's 1925 Pulitzer Prize–winning novel *Arrowsmith*.

But once antibiotics hit the market—sulfonamide in 1935, penicillin in 1942—interest in phages as antibacterial agents mostly dried up (except, interestingly, in the Soviet Union and Eastern Europe).

In 1923, George Eliava, from the country of Georgia, traveled to the Pasteur Institute in Paris, where he met d'Herelle. In 1923, he founded the Eliava Institute in Tbilsi, Georgia, which today remains the epicenter of phage therapy. From 2012 to 2014, more than five thousand patients visited the institute's Phage Therapy Center for treatment. These folks had a variety of bacterial infections, including those caused by antibiotic-resistant bacteria, such as methicillin-resistant *Staphylococcus aureus* (MRSA). They claim that more than 95 percent of these patients showed significant improvement. Today the Phage Therapy Center continues to welcome patients from around the world.

In chapter 15, you read about the problems caused by bacteria that evolve to become antibiotic resistant. But phages also evolve quite quickly. As a result, many bacteria that cannot be killed by antibiotics are vulnerable to attack from phages. (To avert the problem of bacteria becoming resistant to phages during treatment, cocktails of multiple types of phages are often formulated. This strategy is reminiscent of the administration of a combination of antibiotics to treat bacterial infections, such as tuberculosis.)

The first studies of phage therapy in humans that met FDA standards have now been done. So far, the results have been promising. Phages have successfully treated otitis externa caused by *Pseudomonas aerugi-*

nosa, a chronic bacterial ear infection that is notoriously difficult to cure; diarrhea; and infected ulcers associated with leg veins. Properly controlled clinical trials have also been carried out, or are in the planning stage, for the treatment of burn wounds and diabetic foot infections.[2] In June of 2015, the European Medicines Agency (EMA) hosted a workshop on the therapeutic use of bacteriophages. One month later, the National Institutes of Health hosted a similar workshop. That same year, the NIH's National Institute of Allergy and Infectious Diseases announced that phage therapy was one of seven prongs in its plans to combat antibiotic resistance.

Anecdotal reports continue to surface of what appear to be miraculous cures of patients with life-threatening infections caused by antibiotic-resistant bacteria. One such recent case, published in 2017 in the journal *Antimicrobial Agents and Chemotherapy*, describes a sixty-eight-year-old diabetic patient with an overwhelming infection caused by *Acinetobacter baumannii*, which was resistant to all antibiotics.[3] Out of desperation, Dr. Robert Schooley, head of the Infectious Diseases Division at the University of California, San Diego (UCSD), enlisted the help of bacteriophage experts, who tailored a cocktail of phages that were active against this bacterium. The phages were administered intravenously—a first. And they saved the patient's life. (You read about *A. baumannii* in chapter 15. It is one of a growing number of bacteria for which there are no available antibiotics.) This remarkable anecdotal case, coupled with a number of similar cases seen elsewhere, spurred UCSD to launch a clinical center in 2018 to refine phage treatments and help companies bring them to market.

Another remarkable case was reported in 2019. A fifteen-year-old girl with cystic fibrosis also had an antibiotic-resistant *Mycobacterium abscessus* infection, which developed after a double lung transplant.[4] Not only were antibiotics ineffective, but because of the drugs given to her to prevent organ rejection, her immune system was less able to fight the infection. She was given genetically engineered bacteriophages, and she had a full and stunning recovery from the infection.

As of the writing this book, cholera continues to kill more than one hundred thousand people a year. Further, more than 750,000,000 people around the world don't have regular access to safe drinking water. Evidence suggests that phages can someday be used to prevent waterborne infections, such as cholera. (Remember Ernest Hankin's discovery in

1896 of something in the Ganges River that could kill *Yersinia cholerae*.) If so, this would represent a colossal breakthrough in public health.

Perhaps the most promising role of phages in human health involves the things we eat. In 2006, the FDA and U.S. Department of Agriculture approved several phage products for the treatment of foods. The company Intralytix markets two products for the prevention of food-borne infections: ListShield, a phage cocktail that is sprayed on food to kill *Listeria monocytogenes*, and EcoShield, phages that are sprayed on red meat before grinding it into hamburger, to kill *E. coli*. A third product, Salmo-Fresh, targets *Salmonella* in poultry and other foods; it is currently awaiting FDA approval.

Another promising research avenue involves combining phages with antibiotics to target bacteria that are notoriously difficult to treat, such as *P. aeruginosa*.[5] So far, it appears that such combinations can work.

Given the monumental problem of the emergence of antibiotic resistance, it isn't surprising to see that phage therapy is undergoing a renaissance.

19

THE FUTURE OF VACCINES

"For just a few dollars a dose, vaccines save lives and help reduce poverty. Unlike medical treatment, they provide a lifetime of protection from deadly and debilitating disease. They are safe and effective. They cut healthcare and treatment costs, reduce the number of hospital visits, and ensure healthier children, families and communities."—Seth Berkley, CEO of Gavi, the Vaccine Alliance

WHAT ARE VACCINES AND HOW DO THEY WORK?

"Vaccines are the tugboats of preventive health."—William Foege, chief of the CDC Smallpox Eradication Program

A Brief History of Vaccines: One or Two Down and 1,400 to Go

As Michael Osterholm put it in his book *Deadliest Enemy: Our War against Killer Germs*, "It's hard to overstate the impact of vaccines on our history and our lives." In my view, vaccines are by far the most important advance in all of medical science.

A vaccine is a biological preparation—derived from dead or weakened microbes or their constituents—that provides acquired immunity to a particular infectious disease.

You may recall from chapter 6 that the word *vaccine* was coined in the late eighteenth century by Edward Jenner. The material he used to inoculate eight-year-old Joseph Phipps was obtained from a milkmaid who had cowpox (the word for "cow" in Latin is *vacca*). As Jenner hypothesized, the cowpox vaccination protected Joseph later on from a related virus that causes smallpox.

With that one clinical experiment, the era of vaccination was ushered in—roughly three quarters of a century before the discovery that germs cause disease, and a century before the existence of viruses was even conceived.

In retrospect, smallpox was a good place for vaccination to start, because over the course of human history, it has killed more people than all wars combined. In the twentieth century alone, variola major virus, the cause of the disease, killed between three hundred and five hundred *million* people worldwide.

So when, in 1980, the World Health Organization declared smallpox eradicated as a result of intensive vaccination efforts around the world, this was reason for resounding celebration. Indeed, I consider this *the* single biggest achievement in medical history.

To this day, smallpox remains the only infectious disease of humans that has been eradicated from our planet. This landmark achievement required a coordinated effort led by the WHO and many public health pioneers, including the epidemiologists Bill Foege and D. A. Henderson.

The good news is that another viral scourge—poliomyelitis, commonly known as polio—is also projected to exit Earth sometime soon, also as a result of a major global vaccination initiative. Spearheaded by a public–private partnership, the Global Polio Eradication Initiative (involving the WHO, UNICEF, and the Rotary Foundation), the number of cases of polio has plummeted. In 1988, an estimated 350,000 cases of polio were recorded in more than 125 countries. In 2015, there were only seventy-four cases reported in two countries—over a 99.9 percent reduction. And even though there was a slight uptick of cases in 2018, the eradication of polio may be on the horizon.

Other advances in the field of vaccination are detailed in John Rhode's excellent book *The End of Plagues: The Global Battle against Infectious Disease.*[1] As you'll discover if you read it, a "Who's Who" of eminent physicians, scientists, and public health leaders has contributed to vaccine breakthroughs, many of whom were awarded the Nobel Prize for Physiology or Medicine.

But one scientific giant who *didn't* receive a Nobel Prize, because he wasn't alive when the awards program was launched in 1901, was the

French scientist Louis Pasteur. Not only was Pasteur one of the founders of the germ theory of disease, but he also demonstrated the mechanism behind the protective effect of vaccination: the stimulation of the immune system.[2]

The terms *vaccination* and *immunization* are often used interchangeably. Scientifically, though, they mean different things. According to the Centers for Disease Control and Prevention, *vaccination* refers to the act of administering a vaccine into the body to produce immunity to a specific disease; *immunization* is the process by which a person becomes protected against a disease through vaccination. As the result of a successful vaccination, you become *immune*.

As we've seen, to date only one (or, soon, perhaps two) of the estimated 1,400 infectious diseases that can sicken humans have been eradicated by vaccination. This means we still have a very long way to go. While the WHO lists twenty-five vaccines that protect against common infections, for the twenty emerging pathogens highlighted in the "Mortal Enemies" section of this book, vaccines are available that prevent only two: influenza and dengue. And both of these vaccines need to be improved.

But the ongoing need for more and better vaccines shouldn't overshadow the extraordinary benefits of the ones that are currently available. The CDC estimates that vaccines given to infants and young children over the past two decades will prevent 322 million illnesses, 21 million hospitalizations, and 732,000 deaths over the course of those people's lifetimes. And the WHO estimates that vaccination against measles alone has saved 17.1 million lives since the year 2000.

Current Vaccines and Recent Triumphs

Whether you're a parent who takes your child to a clinic for routine vaccinations, or an adult who wonders what vaccines *you* routinely need (or should get before traveling to a developing country), chances are you are overwhelmed by the many variations of what's needed and when. I'm a specialist in infectious diseases, and if you were to ask *me* what vaccinations you need before you visit Botswana, or Turkmenistan, or Patagonia, I'd have to answer, "I don't know. Google it." Better yet, check out the CDC's Advisory Committee for Immunization Practices (ACIP) recommendations—they are comprehensive, up-to-date, and provided online for you and your doctor.

In 2018, the standard schedule for the vaccination of Americans ages zero through eighteen included vaccines against nineteen different viruses and bacteria. Fortunately, this doesn't mean nineteen different vaccinations. The vaccines for measles, mumps, and rubella are given together (in the MMR vaccination), and the vaccines for tetanus, diphtheria, and whooping cough are also administered in a single dose (in the Tdap vaccination).

For adults who are nineteen or older, the ACIP targets fewer microbes. Your vaccination schedule will be based on your age, sex, and whether (and when) you received primary vaccinations as a child.

Developments in the field of vaccination occur so often that breakthroughs sometimes go unrecognized by the general public. So here is a quick overview of the key highlights over the past forty years, that is, during the course of my career as an infectious disease specialist:

- Haemophilus influenza type b (Hib vaccine) was licensed in 1985 to prevent what was then the number one cause of bacterial meningitis in the United States. As a result, this disease, which mostly affects children, has almost completely disappeared.
- Vaccines that target *Streptococcus pneumoniae*, now the most important cause of bacterial meningitis and pneumonia in children and adults, dramatically reduced the number of illnesses and deaths from this pathogen. It has been reengineered several times. The currently recommended vaccine, PCV13, is active against thirteen different types of *S. pneumoniae*.
- Varicella zoster vaccine (VZV) was licensed for use in the United States in 1995; as a result, since then the number of cases of chickenpox has plummeted. VZV also prevents herpes zoster, commonly known as shingles, in adults. Following licensure in 2006, the number of cases of this very painful disease, which most commonly afflicts adults over fifty, has markedly declined.
- Two vaccines that prevent common viral infections and cancers caused by their respective viruses were approved: in 1981, hepatitis B vaccine, which prevents hepatitis B and liver cancer, and in 2006, human papilloma virus (HPV) vaccine, which prevents venereal warts and cancer of the cervix, penis, and anus.
- Rotavirus vaccines, aimed at viruses that were the leading cause of severe diarrhea among young children, were licensed in the United

States in 1998. These prevent an estimated 15 to 34 percent of cases of severe diarrhea in the developing world and 37 to 96 percent of cases in the developed world.

- The Global Alliance for Vaccines and Immunization (or Gavi), an international public–private partnership funded by the Bill & Melinda Gates Foundation and donor countries from the developed world, was launched in 2000. Since its inception, nearly *half a billion* children in developing countries have been vaccinated against life-threatening pathogens, and more than seven million deaths in children have been prevented.

- Meningitis cases in Africa dropped from more than 250,000 in 1996 to eighty cases in 2015—the result of the Meningitis Vaccine Project (MVP), a nonprofit founded in 2001 with support from the Bill & Melinda Gates Foundation and Gavi. At a cost of about five cents per dose, MenAfriVac targets *Neisseria meningitides* serogroup A, the cause of annual epidemics of meningitis across much of Africa.

Vaccines in the Future

Why do we have so few effective vaccines? First, we still don't have a thorough understanding of how pathogens and our immune system interact. Second, there are always large practical hurdles to overcome in developing and delivering any vaccine. While most vaccines are administered by inoculation ("shots") that nobody particularly enjoys, some may be taken by mouth, such as the oral polio vaccine, or by the nasal route. In tropical countries, distribution and storage of vaccines are big challenges. Third, in my professional opinion, there hasn't been enough research funding for vaccines, whether from governments, foundations, nonprofits, or pharmaceutical companies.

Here is another crucial question: with all the opportunities—and needs—for new vaccines, how do researchers and funders decide on what gets the most attention and what gets put on a back burner?

Over the past four decades, I've witnessed many attempts to improve existing vaccines, as well as to develop new vaccines for emerging pathogens. Clearly, one key factor is how big and how imminent a threat the microbe is to humanity. Take Ebola and Zika viruses as two of the most recent for examples.

Both Ebola and Zika were declared a "Public Health Emergency of International Concern" (PHEC) by the WHO—Ebola in August 2014 and Zika in February 2016. And both ignited worldwide panic, Ebola because of its high mortality (over 50 percent) and Zika because of its electrifying spread across the globe and the tragic brain damage of newborn infants that it leaves in its wake. Amazingly, by March 2016, the WHO terminated the PHEC status of Ebola. This was achieved by strict public health measures, not by vaccination. But before the end of the epidemic in West Africa, as was mentioned in chapter 7, a promising vaccine was developed that was ready to go when Ebola virus again raised its ugly head in the Democratic Republic of the Congo in 2018 (see chapter 7).

In some cases, decisions about vaccine development also involve the potential return on investment, especially for diseases that mostly affect residents of impoverished countries. Fortunately, in recent years funding by public–private partnerships or nonprofits, such as Gavi and the Bill & Melinda Gates Foundation, have addressed these inequities. Without these resources, millions of lives would be lost every year and the development of vaccines against big killers like malaria and tuberculosis would grind to a screeching halt.

Although this may sound greedy and coldhearted, in fact the economics of commercially developing vaccines are inherently contradictory. By their very nature, vaccines are designed to put their developers out of business, by keeping people from getting sick. (For example, now is not a good time to invest in vaccines against smallpox.)

The same argument could be made if vaccines were developed that target antibiotic-resistant "superbugs." As discussed in chapter 15, the emergence of bacteria that are resistant to most, and in some cases all, currently available antibiotics is considered by many experts as the single biggest infectious disease threat to human life. If vaccines for such diseases became available, the need for new antibiotics would decline along with the market for them. In July 2017, the Bill & Melinda Gates Foundation announced that it has made vaccines the charity's main strategy to combat antibiotic resistance.

The most exciting recent development in the field of vaccines is the Coalition for Epidemic Preparedness Innovations (CEPI), a billion-dollar public–private partnership launched on January 18, 2017, at the World Economic Forum in Davos, Switzerland. The first targets for the CEPI are vaccines against Nipah virus, Middle East respiratory syndrome, and

Lassa fever, which have the potential to cause outbreaks similar in scale to SARS, Ebola, or Zika.

Given the large number of scientists, physicians, and public health advocates involved, I believe that new or more effective vaccines will materialize in the next few years for the so-called Big Three pandemics: HIV/AIDS, tuberculosis, and malaria. For the same reason, I also expect to see new or more effective vaccines that target *Bordetella pertussis* (the cause of whooping cough), *N. meningitides* type B (a frequent cause of meningitis outbreaks among college students and gay men), and group B *Streptococcus* (the most important bacterial pathogen of newborns).

The most urgent priority, however, should be the development of a universal influenza vaccine—that is, a vaccine that protects against all types of influenza viruses. Currently, a new avian flu pandemic is recognized by many experts as the biggest potential global threat to human life and health. And let's not forget that routine seasonal flu continues to kill between thirty and ninety thousand people annually in the United States alone. Thus, the congressional initiative that was announced in 2018 to develop a universal vaccine is an important step in the right direction.

WHY THE CONTROVERSY?

> "The greatest lie ever told is that vaccines are safe and effective."—
> Leonard G. Horowitz, former dentist and self-help author

Do Vaccines Work?

I hope that what you've read so far in this chapter has convinced you that vaccines are highly effective and extraordinarily valuable. (If, like me, you're over sixty-five and were immunized with only three of the nineteen current routine vaccines, it probably didn't take much convincing. Like me, you probably had measles, mumps, German measles, and chickenpox. And, like me, you may have had a friend or relative who contracted polio and lived in an iron lung, or suffered from life-threatening paralysis.)

But as great as the successes of vaccines have been, no vaccine is perfect. Three notable examples are the vaccines against influenza, pertussis (whooping cough), and mumps.

As you'll remember from chapter 9, flu vaccines must be tailored every year to target viral strains that are predicted to cause the next round of seasonal flu epidemics. In some years, scientists' predictions hit the mark; in others, they miss the mark quite a bit.

As for the current vaccine against whooping cough (a combination vaccine that also protects against diphtheria and tetanus), while it is safer than the older vaccine it replaced, it isn't as effective. That's why you increasingly read or hear about outbreaks of whooping cough in the news. Researchers at the CDC recently reported the reason for the reduced effectiveness of this vaccine: the bacterium that causes the disease, *Bordetella pertussis*, has mutated.[3]

You also may have seen recent news reports of mumps outbreaks throughout the entire United States, especially on college campuses.[4] The reason for this upward trend isn't yet clear. Waning immunity and the need for an adult booster vaccine are being discussed by researchers. As the CDC's Cristina Cardemil says, "Even in the best-case scenario where everyone is vaccinated, we're still going to see a number of mumps cases every year." In 2017, vaccine experts advised that people at high risk of catching mumps during an outbreak get a booster dose of the vaccine, even if they've already been vaccinated.

Are Vaccines Safe?

The main ongoing controversy over vaccines isn't about their efficacy. It's about their safety.

As Mark Honigsbaum points out in his 2016 article in the *Lancet*, "Vaccination: A Vexatious History," concerns about the safety of vaccines aren't new.[5] An 1802 print in the British Museum, a copy of which hangs on my office wall, depicts cows sprouting from the heads and arms of people who have been vaccinated. This reflected a fear of contamination with animal matter.

Fear of vaccines can be traced back still further, to the Revolutionary War. George Washington recognized smallpox as an invisible killer with greater potential as a threat than "the Sword of the Enemy." While he wholeheartedly believed in the efficacy of inoculation, in May 1776 he ordered that no one in his army be inoculated because of the potential side effects of mild illness. He wanted all of his soldiers to be healthy and battle ready. (For an excellent review of the history of the antivaccination movement, I recommend Shawn Otto's book *The War on Science: Who's Waging It, Why It Matters, What We Can Do about It.*)[6]

Just as no public health authority would claim that vaccines *always* work, neither would they claim that vaccines never have side effects. Some people do experience them—though they are not especially common, and when they do occur they are generally mild. These side effects can include fever, pain around the injection site, and muscle pains. In addition, some people may be allergic to a specific ingredient in a vaccine, such as egg protein for vaccines derived from chicken eggs. Severe side effects are extremely rare. However, if you have a compromised immune system, tell your doctor, because a live virus vaccine can result in a life-threatening infection if it is inadvertently given to you—or anyone with a weakened immune response.

In its "History of Vaccine Safety," the CDC explains, "Before vaccines are approved by the Food and Drug Administration, they are tested extensively by scientists to ensure they are effective and safe. . . . The benefits of vaccines far outweigh the risks." While this is certainly true in general, if your immune system puts you at risk for potentially serious side effects, you may be one of the relatively few people for whom certain vaccinations may be unwise.

In the 1970s and 1980s, successful lawsuits against vaccine manufacturers, often without bona fide scientific evidence, halted their production by some drug companies. In response, to reduce liability and respond to public health concerns, the U.S. Congress passed the National Childhood Vaccine Injury Act (NCVIA) in 1986.[7] As a result of the NCVIA and a subsequent Supreme Court ruling, lawsuits against drugmakers over serious side effects from childhood vaccines are prohibited by federal law. This has fueled the anger of antivaccination groups, which take a stand for what they call "health freedom." According to the website of the Health Freedom Coalition, they advocate for a more acceptable balance between individual rights to self-determination and free choice and government concerns for public safety.

The legal requirements regarding the vaccines that are mandatory for children entering day care or school vary from state to state. They can easily be found online, and provisions for medical exemptions and conscientious exemptions are also spelled out.

What makes the difference between the individual "right" to not wear a helmet when riding a motorcycle, or the right to smoke tobacco, and the right to opt out of a mandatory vaccine? For starters, emphysema and head injuries are not contagious—but diseases such as measles and whooping cough are.

This is a significant public health concern. Not getting vaccinated against a contagious infection not only puts you at risk, but it increases the risk for others, who might catch the illness from you. This is of particular concern for people with compromised immune systems, who can't protect themselves by getting vaccinated. If these people do get infected, they are more likely to die or become seriously ill.

What Is Community Immunity—and Why Does It Matter?

Community immunity, also known as herd immunity, is the general immunity of a group of people (or animals) to a pathogen based on the acquired immunity of a high proportion of its members. Because of community immunity, when you get vaccinated against a contagious disease, you benefit not only yourself but others around you who aren't immune.

According to the NIH, "When a critical portion of a community is immunized against a contagious disease, most members of the community are protected against the disease because there is little opportunity for an outbreak. Even those who are not eligible for certain vaccines— such as infants, pregnant women, or immunocompromised individuals— get some protection because the spread of contagious disease is contained."

So when you get vaccinated against a contagious disease, you're protecting not only yourself but also those in the community who are vulnerable to the disease.

There's a flip side to this. If you *don't* get yourself or your child immunized, not only are you or your child at risk, but you add to the risk of other community members.

Measles is the contagious disease that has stirred up most of the controversy over vaccination, partly because, in recent years, several community outbreaks of measles have led to legislative action. One of the most publicized of these was a multistate outbreak that began in Disneyland in Anaheim, California, in 2014. By early 2015, public health officials had identified 125 people with measles. Most of these patients weren't adequately vaccinated against measles. As a result, the herd immunity wasn't strong enough to protect those susceptible to infection.

This California measles outbreak precipitated strict legislation requiring childhood vaccination against measles. (Similar legislation is in place in several other states.) Beginning with the 2016–2017 school year, children whose parents refuse to get them vaccinated must be homeschooled, unless they have a medical exemption (for example, because of a weakened immune system). The law applies to public and private schools, as well as day care facilities.

California's strict vaccination law was applauded by public health organizations, as well as by pediatricians across the country. But it also provoked a major hue and cry from antivaccination activists—often disparagingly referred to as antivaxers—who are lobbying hard in both California and Washington, DC, to revoke legislation that mandates vaccination.

In April 2017, another large measles outbreak erupted, this time in my home state of Minnesota. By mid-July, seventy-nine cases had been identified, twenty-two of whom required hospitalization. (This number of cases already surpassed the total number of measles cases in the entire United States in 2016.)

Like the outbreak in Disneyland, a large majority of patients weren't immunized against measles. The outbreak was first recognized in three children who attended the same day care facility. As was true for these children, most of the subsequent cases were children born in the United States to Somali immigrants—a community that had been targeted by antivaccination activists. (A precipitous drop in MMR vaccination coverage of Somali children preceded the outbreak, demonstrating the impact of the false message of fear spread by such activists.)

Shockingly, in 2019 things are going from bad to worse. By April, 695 cases of measles had been confirmed in the United States—making it the worst year for measles since 1994. Many of these cases occurred in New York City, where Mayor de Blasio declared a health emergency in parts of Brooklyn, requiring vaccinations. In 2000, the WHO had declared that the United States had eliminated measles. But sadly by October 2019, the CDC announced that the country had a reasonable chance of losing its measles elimination status because of ongoing measles outbreaks in New York.

The increase in measles in the United States in 2019 is mirrored elsewhere. In April, the WHO declared that global cases had quadrupled in 2019. Most disheartening to me is that the skyrocketing of measles is occurring soon after the declaration in 2000 that measles had been *eliminated* in the United States. This remarkable step backward is related to reduced community immunity (95 percent immunization coverage is required for protection against measles), which was fueled by antivaccination advocates. And, tragically this is a public health problem with a clear, existing, readily available scientific solution—a measles vaccine that is highly effective and safe.[8]

Whom Do You Trust?

In his 2015 article in the *Journal of the American Medical Association* entitled "Law, Ethics, and Public Health in the Vaccination Debates: Politics of the Measles Outbreak,"[9] Lawrence Gostin underscores how the measles outbreak "reignited a historic controversy about the enduring values of public health, personal choice, and parental rights." Gostin (correctly, in my view) notes that we need to consider the religious, philosophical, and political issues behind the controversy. He concludes that "although vaccine policy is politically divisive, the consensus scientific view is that childhood vaccines are safe and effective, among CDC's 10 great 20th-century achievements and a World Health Organization 'best buy.'"

The side you come down on in the argument about mandatory vaccines depends on how you answer the question, whom do you trust? My personal view, as an infectious disease specialist, is akin to Gostin's. At the end of the day, I trust the scientific evidence. But like other medical doctors, I also share the view that our first priority to patients is *Primum non nocere* (Do no harm). Thus, I worry a lot about side effects of all treatments, including vaccines.

I also understand the skepticism that seems to drive some antivaccination groups. After all, skepticism is a fundamental principal of science. It also underlies one of the 10 Rules of Internal Medicine that I provide in my book *Get Inside Your Doctor's Head: Ten Commonsense Rules for Making Better Decisions about Medical Care*.[10] Rule 6 in this book is "Never trust anybody completely, especially purveyors of conventional wisdom." And the value of vaccination, for most people and virtually all doctors, is conventional wisdom.

On the one hand, conventional wisdom is often conventional precisely because it is wise. But on the other, as we've seen, over time conventional wisdom sometimes has a way of being proven dead wrong.

If you feel strongly that personal freedom trumps all other considerations (including community immunity), or if you feel that you can't trust government regulation (or science), then you're likely going to side with antivaccination campaigners. But you would be making a huge mistake—one that could kill you or someone else.

Tragically, some antivaccination activists put their trust in British researcher Andrew Wakefield. Wakefield published a study in the *Lancet* in 1998 describing measles virus in the digestive systems of autistic children who were given the measles/mumps/rubella vaccine.[11] This study suggested a link between the vaccine and autism. Six years later, the coauthors of this study began removing their names from the article when they discovered that Wakefield had been paid by lawyers who planned to sue vaccine manufacturers. Later, the article was officially retracted by the *Lancet*, Britain revoked Wakefield's medical license, and an investigative reporter wrote a series of articles exposing his fraud.

To this day, Wakefield, who now lives in Texas, is responsible for scaring parents about the unestablished link between measles and autism. It's little wonder that Texas is now the center of the antivaccination movement. (Not coincidentally, the recent measles outbreak in the Somali community in Minnesota mentioned above was preceded by Wakefield's visit and his talk to parents in this community.) In 2017, the CDC reported that, thanks largely to antivaxers, measles cases are on the rise in several states. In Europe, measles cases more than tripled in 2017, and antivaccine activists played a big role in this sad development.

The possible link between the measles vaccine and autism has been extensively examined and reexamined. The evidence clearly doesn't support a causative role. This doesn't mean that there isn't a correlation between the incidence of autism and vaccination rates. But in this case *correlation definitely doesn't mean causation*. In other words, there is absolutely no evidence that vaccines *cause* autism. On the other hand, several studies have found a statistically significant correlation between exposure to particulate air pollution during pregnancy and the development of autism in childhood. And a study from Denmark points to the immediate postpregnancy period as another vulnerable time for the development of air-pollution-related autism.[12] As you would imagine, the ongoing argument about the dangers of vaccines continues to confound the medical community. In her 2017 *Science* article "Four Vaccine Myths," Lindzi Wessel emphasizes the most dangerous false claims about vaccines: (1) vaccination can cause autism, (2) mercury in vaccines acts as a neurotoxin, (3) countering mercury from vaccines can make children better, and (4) spreading out vaccines can be safer for kids. [13]

But at the same time that the science has become crystal clear, it is also clear that facts alone won't convince people to vaccinate their kids. What we need, I believe, is a breakthrough in the science and art of persuasion. And we need this breakthrough now.

Sadly, the state of vaccine confidence seems to be declining. This conclusion is supported by the fact that the number of measles cases has grown worldwide. And in the United States, which recorded a record number of influenza-related hospitalizations during the 2017–2018 flu, vaccination against the flu was only 37.1 percent. Given their role in fostering vaccine hesitancy, it's no wonder that the WHO identified anti-vaxers among the top ten health threats for 2019. But stopping the spread of erroneous claims won't be easy. As with the spread of much other misinformation, social media plays an increasingly devious role. [14]

Vaccines and the Global Community

Despite continuing concerns about mounting cases of measles in Europe and the United States, there's room for optimism. In 2015, the journal *Science* highlighted the work of Steve Cochi, a senior adviser on global immunization at the CDC. [15] Cochi is leading a campaign to eradicate measles. Such a campaign could work; after all, there is an effective vaccine. And, importantly, there is no animal species in which the measles virus could hide out. Just think what an amazing achievement this would be. Eradication of measles would save the lives of more than one hundred thousand children annually. It would also be the most effective way to end the controversy over the measles vaccine. After all, once the disease has been wiped off the earth, no one will ever need to be vaccinated against it.

As James Colgrove pointed out in the *New England Journal of Medicine* in 2016, both persuasion and coercion are necessary to control contagious diseases such as measles. [16] And he reminds us that the central challenge for vaccine policymakers is "ensuring a stable, accessible supply of vaccines for everyone who needs them."

20

MICROBES AND THE SIXTH EXTINCTION

"Extinction is the rule. Survival is the exception."—Carl Sagan

GERMS IN THE DRIVER'S SEAT

A Brief History of Evolution

First, let's consider what the term *evolution* refers to. Biological evolution is the process through which many different kinds of living organisms developed and diversified from earlier forms of life. Contrary to popular belief, this idea didn't originate with Darwin. It dates back to the ancient Greeks. But Darwin is credited with discovering the mechanism behind evolution: natural selection. He formulated the explanation in his highly influential and controversial book *On the Origin of Species by Means of Natural Selection*, published in 1859.[1]

The essence of Darwin's theory of evolution is that all life-forms are related and descended from a common ancestor. The governing force behind the appearance (emergence) or disappearance (extinction) of species is the process of natural selection.

Underlying this process is competition for nutrients and other elements of a supportive environment. According to Darwin, the struggle for existence rewards species that are fit for—that is, well suited to—their environment and eliminates those that aren't. This is the so-called survival of the fittest, though Darwin himself did not use that phrase.

Of course, Darwin's understanding of the microbial world was limited. He took an aquatic microscope along on his famous voyage on the *Beagle* in 1831–1836, and he used a low-powered microscope to examine barnacles and plants. But in formulating his theory, he left microscopic creatures completely out of the picture.

You'll recall from chapter 1 that life on Earth began almost four billion years ago in a hostile environment inhabited by a single-celled microbe dubbed LUCA (last universal common ancestor). From LUCA sprouted two of the three domains of the Tree of Life: Bacteria and Archaea. The third domain, Eukaryota, evolved out of an intimate partnership between bacteria and archaea.

How could a tiny ancient ancestor such as LUCA give rise to the incredible diversity of life-forms we know today? Trying to fathom this mystery requires an appreciation of "deep time."

According to cosmologists' Big Bang theory, our universe was created some 13.75 billion years ago. In comparison, Earth—the only astronomical object currently known to harbor life—is relatively young: about 4.54 billion years old.

The first life-form, LUCA, emerged about 3.8 billion years ago, followed around 1.8 billion years later by the first eukaryotic cells, with their nuclei and other internal organelles. About 1.1 billion years after that—700 million years ago—land plants appeared, followed by animal life about 540 million years ago. *Homo sapiens* showed up around 300,000 years ago. (The oldest known remains of *Homo sapiens* were recently discovered in Morocco by a team of paleoanthropologists from the Max Planck Institute in Liepzig, Germany.)[2]

To comprehend the amount of time it took to get us here, consider this. If the history of life on our planet were compressed to a single twenty-four-hour day, modern humans wouldn't appear until the very last minute. And if we set the clock back to the birth of the universe, our existence would take but a blink of the eye.

Ever since the founding of the discipline of taxonomy by eighteenth-century Swedish scientist Carolus Linneaus, biologists have classified organisms according to their species. One way to define the term *species* is as a group of individual organisms that actually or potentially can interbreed in nature.

Today, most definitions of the term *species* incorporate a genetic understanding of evolution. For example, the paleontologist Anthony

Barnosky defines a species as "a group of plants or animals that can pass their genes on to their offspring, which can, in turn, pass their genes down the line to their offspring."[3]

Barnosky's definition of species applies not only to organisms that pass their genes on to one another sexually, such as plants and animals, but also to those that reproduce asexually, such as bacteria and archaea, that simply divide by binary fission. And if you are willing to buy that viruses belong in the Tree of Life, even these tiniest of microbes pass on their genes to their progeny, after first commandeering host cells.

Linneaus and Darwin didn't know a thing about genes. If scientific communication back then had been better, Darwin might have learned of the pioneering work of an Augustinian monk, Gregor Johann Mendel, the father of modern genetics. His ingenious experiments hybridizing peas in 1856–1863 led to an understanding of heredity. He presented his electrifying findings in 1865 at a small meeting in Brno (now in the Czech Republic), but for the next thirty-five years his work was mostly ignored and unrecognized.

In 1900, sixteen years after his death, Mendel's work was rediscovered by three different scientists. In 1909, Danish botanist Wilhelm Johanssen coined the term *gene* to describe Mendel's units of heredity. And many years and Nobel Prizes later, the nature of genes—or, more technically, DNA or deoxyribonucleic acid—was eventually unraveled.

Another profound development in the field of evolutionary biology came in the late twentieth century, when Carl Woese and his colleagues started what might be called the Metagenomics Revolution. As described in earlier chapters of this book, these scientists used modern molecular biology tools to identify microbes that can't be grown (or cultured) in the laboratory. They and other, later researchers discovered genetic material from a staggering number of microbial species in the environment—more than 99 percent of which we didn't even know existed.

How many different species are there on Earth that we know about? How many are there likely to be that we don't yet know about? And how many have been here and gone extinct? Sit down, take a deep breath, and get ready for some seriously large numbers.

- According to the International Union for the Conservation of Nature, there are an estimated 8.7 million species now alive on planet Earth.[4] (This number does *not* include microbes; more on these below.) The IUCN guesses that less than 15 percent of these species have even been discovered. To put this another way, we may be completely unaware of 85 percent of all the complex creatures on our planet—and a much higher percentage of the simpler, microbial ones.

- Some experts suggest even higher numbers. They believe that Earth harbors 10 to 50 million species of animals, of which 3 to 30 million (97 percent) are invertebrate species and one million are insect species. They also believe that there are 300,000 to 400,000 species of plants (250,000 of which are flowering), and 6,100 species of amphibians.

- As for microbes, scientists estimate that there exist 10 *quintillion* (or 10^{19}) species of bacteria (although nobody knows for sure). Of these, only 15,000 species are known to us—and a mere 1,400 species can harm human beings. Scientists estimate that there are 10 million species of fungi. The number of virus species is anyone's guess—somewhere between hundreds of millions to at least a billion.

- Using sophisticated scaling laws and data from a large number of sampling sites, in 2016 Indiana University researchers Kenneth Locey and Jay Lennon estimated that there are as many as one *trillion* different species now alive on Earth.[5] This number includes viruses, whales, and everything in between.

The Big Five Mass Extinctions

Along the trajectory of the past 3.8 billion years, a mindboggling number of species emerged and became extinct. Of the thirty billion animal and plant species that ever existed on planet Earth, more than 99 percent are gone forever.

Geologists, paleontologists, and other earth scientists divide the long history of our 4.54-billion-year-old planet into hierarchical chunks of time. From the longest to the shortest, this hierarchy includes eons (half a billion years or more), eras (several hundred million years), periods (tens to one hundred million years), epochs (tens of millions of years), and ages (millions of years).

In the very first eon, the Hadean—from the Greek word for *Hades*—the surface of Earth must have been like our image of Hell. Of crucial importance for life, however, oceans formed. (Water served as the habitat for all microbes, plants, and animals, until the first land plants cropped up during the Ordovician Period and amphibians crawled out of the sea in the Devonian Period about 370 million years ago.)

It was during the change to the next eon, the Archaean, 3.8 billion years ago, that things got interesting—LUCA, Earth's first living cells, emerged. Sometime thereafter, LUCA gave birth to the domains Archaea and Bacteria. It wasn't until the Proterozoic Eon that the final domain, Eukaryota, emerged about 1.9 billion years ago.

One of the newest species, *Homo sapiens*, didn't appear on the scene until the Pleistocene Epoch, as mentioned earlier, about three hundred thousand years ago. At the end of the last ice age, human civilization entered the Holocene Epoch—the name given to the last 11,700 years of Earth's history.

During this great sojourn of time, billions of animal and plant species arose and died out. Some were outcompeted by other species; others were destroyed in one or more cataclysmic events.

During the Mesozoic Era, dinosaurs ruled the animal world. This time period, known as the Age of the Reptiles, stretched from 240 to 65 million years ago.

As first hypothesized in 1980 by Luis and Walter Alvarez of the University of California, Berkeley, scientific evidence indicates that all dinosaurs, *as well as 76 percent of all of Earth's known species*, were snuffed out by an asteroid that hit the Yucatan Peninsula sixty-five million years ago. The impact of the asteroid threw up enough dust to cause severe climate change. Most species simply could not adapt to the changes, and died out.

Amazingly, this was one of the *smaller* of the so-called Big Five Mass Extinctions.[6] By definition, a mass extinction is an event or period of time when more than 75 percent of Earth's species died off. The Big Five Mass Extinctions, listed below according to their geological periods, represent the largest in an otherwise relatively smooth continuum of extinction events:

Ordovician–Silurian extinction events: 450–440 million years ago.

Late Devonian extinction event: 370–360 million years ago.

Permian–Triassic extinction event: 252 million years ago. This was Earth's largest extinction and is sometimes known as the Great Dying. Between 90 and 96 percent of all species were lost. In the sea, even the highly successful marine arthropods called trilobites were wiped out. This event also ended the primacy of mammal-like reptiles. It took thirty million years for vertebrates to recover.

Triassic–Jurassic extinction event: 201.3 million years ago. Most nondinosaurian archosaurs and large amphibians were eliminated, leaving dinosaurs with little competition on land.

Cretaceous–Paleogene extinction event: 65 million years ago. All nonflying dinosaurs became extinct. Mammals and birds emerged as the dominant large land animals.

What do these mass extinctions of animals and plants have to do with germs? Clearly, a vast number of microbes were also eradicated during these dire times. But nobody knows for sure what microbial species were lost and which proved the fittest. It seems likely, however, that the archaean extremophiles—the microbes that can survive great cold, great heat, great pressure, or other extreme conditions—would have been quite content, no matter how severe conditions were. (It's no wonder that the search for life on other planets has these microbes in mind.)

Germs have clearly played a role in shaping extinctions, just as they have in the evolution of new species. You may recall from chapter 2 that we can thank the phytoplankton cyanobacteria for adding oxygen to Earth's atmosphere about 2.3 billion years ago (a phenomenon referred to as the Great Oxygenation Event).[7] Without this precious gas, aerobes—including us—wouldn't have emerged.

On the negative side of germs, recent studies by Dan Rothman, a geochemist at the Massachusetts Institute of Technology, and his colleagues implicated an archaean microbe, *Methanosarcina*, in the Great Dying 252 million years ago.[8] In this case, it involved another gas: methane. These scientists hypothesize that blooms of *Methanosarcina* in the oceans resulted in the release of large quantities of methane, which traps heat in the atmosphere much more efficiently than does carbon dioxide. Just prior to the mass extinction, a tremendous eruption of volcanoes in what is now Siberia belched forth huge amounts of carbon dioxide. The combination of methane produced by *Methanosarcina* and carbon dioxide from volcanoes appears to have caused global temperatures and ocean acidification levels to rise, precipitating catastrophic climate change. (In fact, global warming appears to have contributed to three of the five mass extinctions.)

Many biologists believe we are now living through a sixth major mass extinction. Global warming is a culprit in this mass extinction as well. But this extinction event is quite different. Unlike the previous five mass extinctions, scientists believe that this one is caused by a single animal species: *Homo sapiens*.

THE SIXTH EXTINCTION

"What is man? Man is a noisome bacillus whom Our Heavenly Father created because he was disappointed in the monkey."—Mark Twain

Brief History of Human Evolution

We humans are bunched together with related species in a family called the Hominidae, which includes gorillas, chimpanzees, bonobos, and orangutans. (The nonhuman Hominidae are collectively known as the great apes.) About six million years ago, humans and chimpanzees diverged from a common grandmother. Genetic studies have shown that the earliest documented members of the genus *Homo* appeared around 2.3 million years ago. They belonged to the species *Homo habilis*.

Paleoanthropologists and archaeologists continue to discover fossils of earlier hominids of the genus *Australopithecus*, from which the genus *Homo* diverged. The most famous of these early human ancestors was a member of the *Australopithecus afarensis* species affectionately named Lucy. Recent studies by University of Texas researchers suggest that she died from injuries sustained after tumbling out of a tree about 3.2 million years ago.[9] According to the recent African ancestry theory, *Homo sapiens* emerged first in Africa around three hundred thousand years ago. We quickly went on the move. Possibly driven by drought or other environmental factors, humans migrated out of the African continent fifty to one hundred thousand years ago, replacing other species: *Homo erectus*, *Homo floresiensis*, and *Homo neanderthalensis*. (The last of these species, the Neanderthals, arose somewhere in Europe and Asia. They occupied the same caves as *Homo sapiens* in the Middle East and Europe. Genetic analysis indicates that there was interbreeding of Neanderthals with our species some thirty-five to eighty-five thousand years ago.)[10]

The fossil record shows that we had lots of company for at least 150,000 years. A measly fifty thousand years ago there were at least three other *Homo* species sharing the planet with us. So what happened to all of our relatives?

In his book *Sapiens*, the Hebrew University historian Yuval Noah Harari suggests that this had something to do with the appearance of new ways of thinking and communication (what he calls a cognitive revolution) some thirty to seventy thousand years ago.[11] He hypothesizes that, among all species of the genus *Homo*, ours is the only one that acquired the ability to think and speak about things that do not exist.

From a Darwinian perspective, this characteristic—imagination and an ability to tell stories—could have come in handy. You can imagine how creative thinking would foster survival and give *Homo sapiens* a competitive edge over other species.

But there are other hypotheses to consider—including some from the perspective of germs. Given their history of shaping human societies, microbes carried by *Homo sapiens* could have played a defining role in culling out competing hominids. In support of this hypothesis, a recent study from the University of Cambridge suggests that Neanderthals across Europe may well have been infected by microbes carried out of Africa by waves of *Homo sapiens*.[12]

Another possibility is that, in contrast to other *Homo* species, *Homo sapiens* may have evolved an immune system that was more effective in defending us against microbial enemies. It is also possible that humans acquired a microbiome that better promoted the health of our species.

The Hologenome Theory of Evolution

The hypothesis that the human microbiome played a role in the evolution of our species is consistent with an emerging perspective in evolutionary biology called the hologenome concept. To understand this concept, we need to return to some of the key ideas in chapter 3.

As you'll recall from that chapter, the human microbiome is the many ecological communities of symbiotic microbes that share our bodily surfaces. (They are called symbiotic because they live together with us—and either they cannot live without us, we cannot live without them, or both.)

Here is the essence of the hologenome concept of evolution: the holobiont (a host organism plus its symbionts—its symbiotic microbes) along with its hologenome (the genes of the host organism plus the genes of its symbionts) shape natural selection. In a 2016 article in *mBio*, Israeli scientists Eugene Rosenberg and Ilana Zilber-Rosenberg write, "A large number of studies have demonstrated that symbionts contribute to the anatomy, physiology, development, innate and adaptive immunity, and behavior and finally also to genetic variation and to origin and evolution of species. Acquisition of microbes and microbial genes is a powerful mechanism for driving the evolution of complexity."[13] In other words, your DNA + the DNA of your microbes = your hologenome. Furthermore, our species, *Homo sapiens*, is perhaps best defined not just by the genome of our cells but by the genome of those cells *plus* the genome of our microbial symbionts. Together with our symbionts, we have been evolving together.

This concept is controversial, but it is compatible with the broader understanding of the inseparable relationship between microbes and humans—and, for that matter, between microbes and all animal and plant ecosystems. And it should be pointed out that the holobiont theory is gaining increasing traction in the field of evolutionary biology.[14] Most of us who buy this idea think primarily of the role of bacterial symbionts in shaping evolution. But we may need to think again. A 2016 article in the journal *eLife* by Stanford University researchers suggests that viruses are a major driver of human evolution.[15] The function of genes is to code for proteins—the building blocks of all cells. The findings by Enard and Petrov suggest that an astonishing 30 percent of all protein adaptations since humans diverged from chimpanzees have been driven by incorporation of genetic elements of viruses into our ancestors' genomes.

The Anthropocene Extinction

According to geologists, we are officially living in the Holocene Epoch that began only 11,700 years ago. But, given mounting evidence of the enormously negative impact that members of *Homo sapiens* have had on Earth, in May 2019 the Anthropocene Working Group agreed to submit a formal proposal to the International Commission on Stratigraphy declaring an epoch-making change in geological terminology. (Good grief! Remember that epochs are supposed to last for tens of millions of years.) They propose that we are now in the Anthropocene Epoch. (The name *Anthropocene* is derived from two Greek words, *anthro* meaning "human" and *cene* meaning "new.") They also propose that we are in the midst of the Sixth Extinction, which they call the Anthropocene Extinction.[16] A collection of researchers known as the Anthropocene Working Group proposed in August 2016 that the post–World War II boom of the late 1940s and 1950s be recognized as the Anthropocene's start date. Their recommendation is now under consideration by the International Commission on Stratigraphy, the bureaucracy that governs the naming of geological time. The current scientific consensus favors the epochal name change.

However, controversy abounds regarding what to use as the official start date of the Anthropocene. Arguably, humans' most disruptive technologies began many thousands of years ago, when our ancestors began planting crops and crossing the globe. But over the subsequent millennia, humans have left their mark on nearly every nook and cranny of planet Earth.[17]

The extinction of species has gone on since life began almost four billion years ago. However, because of human activities, the rate at which species go extinct has accelerated dramatically.

From the accumulated evidence, scientists can calculate the normal "background extinction rate." In the case of the 5,416 mammal species, for example, the background extinction rate predicts that one species naturally disappears about every seven hundred years. In the Anthropocene Epoch, however, mammal species are going extinct sixteen times faster than normal.

Birds are even worse off, with extinction rates of nineteen times the background rate. And amphibian species are disappearing at a mind-boggling *ninety-seven times* the normal background extinction rate. In fact, currently about 30 percent of all the world's nonmicrobe species are threatened with extinction.

Why are amphibians the most endangered class of animals? The answer is, at least partly, a microbe. An emerging fungal infection called chytridiomycosis, caused by a pathogen with an appropriately vile name *Batrochochytrium dendrobatidis*, or *Bd* for short, is decimating many of the world's 6,100 amphibian species. First emerging in Asia in the early twentieth century, it spread quickly throughout the world. Today this fungal infection is found in at least thirty-six countries. According to the website SaveTheFrogs.com, chytridiomycosis may be the worst disease in recorded history in terms of its impact on biodiversity.[18]

Among mammal species, bats are in the biggest trouble. About a quarter of the 2,220 bat species are on extinction watch lists. Once again, microbes are at least partly to blame. As you'll recall from chapter 9, another emerging fungal infection, called white-nose syndrome, is killing bats at an alarming rate in the United States.

The alarming loss of animal and plant species was described in detail in 2019 in the United Nations report by the Intergovernmental Science-Policy Platform on Biodiversity and Ecosystem Services.[19] In it we are told that because of humans, as many as one million plant and animal species are now at risk of extinction, posing a dire threat to ecosystems all over the world.

The reasons behind these extraordinary extinctions of animal and plant species are complex and often not completely understood. But climate change, the destruction of habitats, and the emergence and spread of pathogens all appear to be involved.

Climate Change and Infectious Diseases

As you've read in this book, a remarkable worldwide acceleration of new and reemerging infectious diseases in humans has recently unfolded, beginning in the last quarter of the twentieth century. As you've also read, most of the 140 or so emerging infectious diseases involved human behavior (or misbehavior).

To the extent that humans contribute to global warming or otherwise foul the environment, certain emerging pathogens benefit. The impact of global warming on Earth's microbiome is only beginning to be understood. Hotter temperatures can lead to increased growth and genetic changes of microbes. In an article in the journal *Nature Climate Change* in 2018, Derrick MacFadden and colleagues show that increasing temperatures promote the increased antibiotic resistance of common bacterial pathogens.[20] (You'll recall from chapter 15 that antibiotic resistance is viewed by many experts as our biggest global infectious disease threat.)

The most compelling evidence linking climate change to emerging infections, however, involves diseases that are spread in our water and through insect bites. Due to global warming and increased rainfall, habitats favoring insects that spread disease are expanding.

Climate change is especially warmly welcomed by mosquitoes and ticks. Mosquitoes spread dengue, chikungunya, Zika virus, malaria, and (with the help of birds) West Nile virus. Ticks spread the infectious agents that cause Lyme disease, anaplasmosis, and babesiosis.

As for waterborne infections, cholera still kills about one hundred thousand people a year. Five times that many die each year from malaria, which is fostered by warming temperatures and increased rainfall in some areas due to climate change. (In the recent book *Climate Change and the Health of Nations: Famines, Fevers, and the Fate of Populations*, Anthony McMichael and his colleagues make a convincing case that climate change played a role in these, as well as several other, catastrophic epidemics.)[21]

In the past several years, we have also witnessed the emergence of several extraordinarily large, harmful, offensive algal blooms—off the coasts of Florida, in Lake Erie, and even near the Greenland ice sheet. Scientists recently found that the ocean has been increasing its production of algae by some 47 percent since 1997. Algal blooms not only look and smell bad, but they have severe impacts on human health, aquatic ecosystems, and the economy.

Algal blooms are formed by cyanobacteria (blue-green algae), which, as you've read, are photosynthetic bacteria, like those that oxygenated Earth's atmosphere. While algae don't directly infect humans, some algae species produce toxins that can sicken us. Because algae flourish with climate change, more and bigger algal blooms are occurring.

Microbes to the Rescue?

Many climatologists warn that the extinction of some amphibian species should be considered the proverbial "canary in the coal mine," warning us of serious potential danger to all animals, including us, as well as to plants, 37 percent of species of which are already endangered.

On a hopeful note, however, remember Professor Yuval Harari's hypothesis as to why our species of *Homo* survived: the ability to think imaginatively and creatively. Although it may have taken *Homo sapiens* way too long to get the message regarding the threat of climate change (and some members of our species still haven't gotten it), the magnitude of the challenge is now recognized by most governments.

Aimed at curbing emissions of greenhouse gases to keep global warming well below 2 degrees Celsius above preindustrial temperatures, the historic December 2015 Paris Agreement (developed within the United

Nations Framework Convention on Climate Change) was approved by representatives from 195 countries. By September 2016, 180 countries had signed the agreement, and twenty-six had ratified it, including the United States and China. While this extraordinary degree of international cooperation was applauded by virtually all countries, the current American administration, under president Donald Trump, pulled the United States out of this agreement in May 2017. In fact, by the end of 2017, the United States was the *only* country in the world not a partner to the Paris Agreement.

There are other reasons to be optimistic about *Homo sapiens* rising to the occasion and averting an impending Anthropocene Extinction. One of the benefits of mitigating the impact of global warming is helping put the brakes on emerging infectious diseases. A positive, related step in this direction was the recent development of a coordinating body to manage dangerous microbes: the international, independent, multistakeholder Commission on Creating a Global Health Risk Framework for the Future, which recognizes infectious diseases as one of the biggest risks facing humankind—right up there with wars and natural disasters in their capacity to endanger human life, health, and society. The commission's goal is to recommend an effective global architecture for recognizing and mitigating the threat of epidemic infectious diseases.

On another positive note, a growing number of scientific leaders and organizations are promoting a One Health approach to the challenge of infectious illnesses. You'll remember from chapter 5 that the concept of One Health is simple: we are all in this together. Solutions to problems affecting our world thus involve networks of healthcare and public health professionals, as well as politicians, ethicists, lawyers, economists, climatologists, geologists, microbiologists, anthropologists, religious leaders, and many other professionals.

Reflecting an even broader vision is an allied initiative called Planetary Health, which is now gaining momentum. Championed by the medical journal the *Lancet*, Planetary Health's concern is the care, health, and wellness of "Patient Earth."

Ironically, one of the most promising and innovative approaches to curtailing an Anthropocene Extinction involves germs—ones that are our intimate friends. These include not only the friendly microbes in our microbiome but also those that are friends of the earth.

As you now know, our ancient microbial ancestors have had billions of years of experience as geoengineers and biogeochemists. Some bacteria, for example, think plastic is delicious. Others happily consume oil spill gases, radioactive contamination, nylon, sulfur, paper, and pollutants. Still other microbes produce oxygen or suck up vast amounts of carbon dioxide as they grow, making the atmosphere cooler. Some evidence suggests that bacteria can play a role in the development of sustainable energy. And in a full-page ad in the *New York Times* in the fall of 2018, the fossil fuel giant ExxonMobil extolled algae as an unexpected ally in our energy future.

And let's not forget fungi. In the January 2018 issue of *Frontiers in Microbiology*, Michael Daly, professor of pathology at the Uniformed Services University of Health Sciences in Bethesda, Maryland, and his colleagues reported the extraordinary potential of the microscopic fungus *Rhodotorula taiwanensis*.[22] This humble fungus may help us clean up huge amounts of radioactive waste in Earth's soil and groundwater. Meanwhile, the Danish biotechnology company Novozymes is working to fight climate change by developing enzymes like those found in oyster mushrooms. These enzymes clean clothes just as effectively as current commercial laundry detergents but at lower temperatures. The energy savings could be significant, since washing machines account for 6 percent of all household energy use in Europe.

There is also good microbe-related news regarding Earth's largest environment, its oceans. In 2015, the Tara Oceans Expedition announced an interdisciplinary research program to define the structure and function of the ocean microbiome. This program seems particularly fitting because oceans spawned and supported life for 3.8 billion years of Earth's history. Also, oceans absorb about 90 percent of the heat building up from the release of greenhouse gases. While the cataloguing of the members of the ocean's microbiome doesn't represent a solution to problems like climate change, it is an important first step.

Given the many recent scientific breakthroughs, especially those in microbiome science, it appears we have a fighting chance of averting the Sixth Mass Extinction. If not, the birds have a good shot at continuing to make it, as they did in the Cretaceous–Tertiary extinction. But one thing is for certain: germs will survive.

21

SCIENCE, IGNORANCE, AND MYSTERY

"Science is not only compatible with spirituality, it is a profound source of spirituality. When we recognize our place in an immensity of light-years and in the passage of ages, when we grasp the intricacy, beauty, and subtlety of life, then that soaring feeling, that sense of elation and humility combined, is spiritual. . . . The notion that science and spirituality are somehow mutually exclusive does a disservice to both."—Carl Sagan

As you read in chapter 20, climate change is arguably the most important looming threat to human health, as well as to the health of Patient Earth. While humans are the root cause, microbial conspirators are doing their part to wreak havoc on many species, including our own.

As I completed this book, two books were published that lifted my spirits against the two biggest doom-and-gloom scenarios—climate change and influenza pandemics. I recommend both books highly: Hans Rosling's *Factfulness: Ten Reasons We're Wrong about the World—and Why Things Are Better Than You Think*[1] and Stephen Pinker's *Enlightenment Now: The Case for Reason, Science, Humanism, and Progress*.[2] Rosling and Pinker provide convincing evidence that in the last half century, scientific and technological advances have resulted in extraordinary—and often poorly appreciated—improvements in almost all sociopolitical and public health spheres.

As you read in the previous chapter, clearly there is reason for hope that climate change can be halted. And that hope is science. Science is also the key to solving the unanswered questions and challenges of the emerging infectious diseases that you've read about in this book.

Throughout this book, you've seen how scientists have solved problems, some of which seemed insurmountable at the time. What, then, are the key characteristics of science—and of individual scientists—that will enable us to tackle climate change, as well as the myriad challenges raised in the "Mortal Enemies" section of this book?

Two major motivating forces of all good scientists are curiosity and the related quality of imagination. Most nonscientists don't realize that imagination is, and always has been, one of the cornerstones of science. As Albert Einstein observed, "I am enough of an artist to draw freely upon my imagination. Imagination is more important than knowledge. Knowledge is limited. Imagination encircles the world."

Throughout history, many (perhaps even all) of the extraordinary scientific discoveries regarding germs were made by scientists who thought outside the box. Initially, their ideas were considered outrageous and were ridiculed. Eventually, however, they were proven right—and prescient.

One of the best examples of this is the scientist who first recognized germs, Antonie van Leeuwenhoek. His original description in 1683 of microbes ("very small animalcules, very prettily a-moving"), which he saw with the microscope he invented, was viewed with great skepticism and derision by members of the esteemed Royal Society of London. It wasn't until his experiments were reproduced that they were accepted. (Little did van Leeuwenhoek know that he was destined to become the father of microbiology.)

This anecdote also underscores two other key aspects of science. The first is the requirement that all hypotheses be tested in experiments and that other researchers be able to reproduce the same results. The second is the sometimes overlooked importance of technology. If van Leeuwenhoek had not invented the microscope, his ideas would not have been proven through observation—and they would likely have gone nowhere for many years.

In a similar but more contemporary vein, consider the work of Carl Woese. Almost three centuries after van Leeuwenhoek, in 1977 Woese shook the world of biology with the discovery of Archaea, a previously unknown domain of the Tree of Life. This finding was made possible because of a new technology called metagenomics: the direct genetic analysis of genomes in environmental samples. Woese and his colleagues led the way to the discovery that our bodies, and our planet, are inhabited by a truly astronomical number of microbial species, the vast majority of which are beneficial or harmless. As a result, as you read in chapter 20, scientists are now imagining ways to harness microbes to save humanity.

As I write this chapter, the number of promising new innovations that can nurture our intimate friends or combat our mortal enemies seems almost limitless. One of the most exciting of these breakthroughs that set the entire scientific world abuzz was the announcement of a genome-editing tool called CRISPR-Cas9. (*CRISPR*—pronounced "crisper"—stands for clustered regularly interspaced short palindromic repeats. *Cas* refers to CRISPR-associated proteins.) With this tool, scientists can edit one or more genes in the genome of any cell in any animal or plant.[3] The discoverers are undoubtedly on the fast track to a Nobel Prize.

CRISPR-Cas9 works in the exact same way that bacteria and archaea defend themselves against their natural enemies, bacteriophages. This means that, although humans may have discovered the technology that makes CRISPR-Cas9 work, microbes actually invented it. (And, true to form, three studies that originally proposed this idea for creating a genome-editing tool were submitted more than a decade ago to high-profile scientific journals—and all three were rejected.)

CRISPR-Cas9 has enormous potential. Using genome editing, it may be possible to treat a range of medical conditions that have a genetic component, including cancer, hemophilia, sickle cell anemia, and devastating genetic diseases such as muscular dystrophy and cystic fibrosis. But CRISPR-Cas9 gene editing technology is only in its infancy, and many questions remain about its safety and efficacy—and, in certain contexts, its morality. The announcement by the Chinese scientist He Jiankul on November 28, 2018, that he had created the world's first gene-edited babies—twin girls who had a gene altered to make them resistant to HIV infection—was quickly renounced by the scientific and bioethics communities.[4]

This type of revolutionary gene-driven technology also affords us, for the first time, the power to alter or eliminate entire populations of pests in the wild, such as particularly nasty species of mosquitoes, as you read in chapter 8. (But again, before such technology is widely implemented, many ethical issues need to be resolved.)

Some further good news: In March 2018, a group of Israeli scientists working at the Weizman Institute of Science reported a major breakthrough in the journal *Science*. They discovered a large group of previously unknown immune systems in bacteria and archaea that may be able to protect human beings against viruses.[5]

Another extraordinary technological breakthrough that is likely to advance our fight against microbial enemies as well as enhance the search for microbial allies is development of artificial intelligence (AI). According to John Bohannon in a July 2017 article in *Science*, robots working for the biotechnology company Zymogen "spend their days carrying out experiments on microbes, searching for ways to increase the production of useful chemicals." And AI isn't just a tool. "In some labs, it conceives and carries out experiments—and then interprets the results."[6] (How's that for imagination!)

I've highlighted the roles of curiosity and imagination in advancing science, but an effective scientist also needs something else: doubt, a vital counterbalance to imagination. Doubt is especially necessary when a new observation or theory emerges that challenges conventional wisdom.

At the same time, however, too much doubt—or doubt that is misapplied—can sometimes impede scientific progress. Consider the experience of one of the founders of the germ theory of disease, Louis Pasteur. In the 1860s, Pasteur proposed that bacteria are responsible for the decay of living matter, as well as for the process of fermentation. At the time, however, the prevailing theory—and the conventional wisdom—was that life could spring from nonliving things, through a process called spontaneous generation. Purveyors of conventional wisdom doubted Pasteur's ideas at first. But as evidence in favor of those ideas mounted, many of Pasteur's colleagues clung to their doubt and vehemently opposed his ideas and observations. As a result, Pasteur was forced to spend decades, rather than months or years, providing scientific evidence that finally debunked the entrenched theory of spontaneous generation once and for all.

Doubt is very different from denial—the refusal to even entertain an idea that's strange or new, or to accept an idea even after it is supported by compelling evidence.

Properly employed, doubt makes us careful and thorough, not dismissive or arrogant. But denialism—the use of false arguments that sound legitimate but are in fact entirely bogus—only impedes scientific progress. Sometimes it even kills people. Consider, for example, the to-bacco industry's denial of the link between cigarette smoking and lung cancer, heart disease, and other illnesses.

When denialism is used to prop up conventional wisdom in the face of compelling evidence to the contrary, the entire human race loses. Some-times, so does our planet.

One more thought on conventional wisdom. A large majority of clima-tologists now agree that the earth is experiencing a cataclysmic disrup-tion: significant worldwide climate change. Their proposition, built on sophisticated techniques of climate modeling, has risen to the level of conventional wisdom. This conventional wisdom makes great sense to me and many, many other scientists.

Yet conventional wisdom needs to be debated and questioned because sometimes it turns out to be dead wrong. (In this case, while I hope climatologists are wrong, let's not bet the lives of future generations of people, other animals, and plants on their being in error.)

But one thing we do know for certain is that germs have profoundly shaped the history of life on our planet. And one prediction you can bet on—based on the startling eruption of mortal enemies in the past half century, germs will wreak havoc in the future. Thus, even though we don't know for sure which ones will emerge or when, the recent work being done internationally to prepare us and the planet for them is highly encouraging.

WHAT WE DON'T KNOW

The joy of discovering new knowledge about the natural world is one of the things that drives most scientists. But science also requires an ongo-ing, and humbling, admission of ignorance. As Lewis Thomas, the emi-nent physician and author of many books and essays about the natural world, sagely observed, "The greatest of all the accomplishments of 20th century science has been the discovery of human ignorance."

Thomas's view is expanded on in Columbia University neuroscientist Stuart Firestein's wonderful book *Ignorance: How It Drives Science*.[7] Firestein stresses that "the recognition of ignorance is the beginning of scientific discourse. When we admit that something is unknown and inex-plicable, then we admit it is worthy of investigation."

The biggest unknown in the story of life is its origin. Where on Earth, or elsewhere, did our last universal common ancestor, LUCA, come from? Despite all our advances and all our knowledge, we still have no idea.

Nick Lane, an evolutionary biochemist at the University of London, sums up our profound ignorance about how complex cells arose in his book *The Vital Question: Energy, Evolution, and the Origins of Complex Life*. He writes, "There are no surviving evolutionary intermediates, no "missing links" to give any indication of how or why these complex traits arose, just an unexplained void between the morphological simplicity of bacteria and the awesome complexity of everything else. An evolutionary black hole."[8] Antonio Damasio, the neuroscientist you read about in chapter 2, makes a similarly strong case for scientific humility:

> It is apparent that we talk with some confidence about the traits and operations of living organisms and of their evolution and that we can locate the beginnings of the respective universe about thirteen billion years ago. We do not have, however, any satisfactory scientific account of the origins and meaning of the universe, in brief, no theory of everything that concerns us. This is a sobering reminder of how modest and tentative our efforts are and of how open we need to be as we confront what we do not know.[9]

Many centuries before the scientific revolution in sixteenth- and seventeenth-century Europe, two Chinese scholars recognized the limits of human knowledge. Confucius said, "Real knowledge is to know the extent of one's ignorance." And his sixth-century BCE contemporary Lao Tzu professed, "Those who have knowledge don't predict. Those who predict don't have knowledge."

Some scientists believe that human uncertainty in predicting the future can be surmounted. They insist that, given enough time, experience, and information, we can know everything about the natural world. Other scientists aren't so optimistic. They see some aspects of the future as inherently unknowable—what is called an aleatory dilemma.

As you've read in this book, one of the most serious areas of scientific ignorance is the inability of experts to accurately predict the future. To give just one example, nobody saw coming *any* of the emerging infections highlighted in the "Mortal Enemies" section of this book. Although many scientists are working feverishly to improve on predicting the "what, where, and when" of our mortal enemies, these look like what Firestein refers to as "unknowable unknowns." (By the way, the inability to predict accurately isn't limited to science. Consider for a moment the stock market, or most momentous world events.)

Our ignorance doesn't just involve the big challenges, like predicting which mortal enemies will emerge next, and if or when the enemies we are now battling will die out. In fact, as you've read, glaring new areas of ignorance about microbes regularly continue to appear. For example, of the staggering number of microbes in our microbiome, which ones are important for our health, and which play a role in certain diseases? Why is it that only a small percentage of people who get infected with some mortal enemies, such as West Nile virus, dengue, and Zika, get sick? Meanwhile, why is it that everyone who has the misfortune of acquiring an HIV infection ultimately gets sick, and—unless they get treated—will die? Right now, the emphatic answer to each of these questions is we don't know.

As you've also read, the nature of many germs remains profoundly mysterious. We don't know for sure, in many instances, just how they do what they do.

Fortunately, a great many scientists are eager to find out. In the words of Stuart Firestein, "Being a scientist requires having faith in uncertainty, finding pleasure in mystery, and learning to cultivate doubt." (And it may be added that the only pleasure that surpasses a sense of mystery is the occasional experience of solving one.)

At the same time that scientists are probing the profound mystery of how life began, others are worrying about the extinction of our species (along with many other forms of life). Intriguingly, these concerns have spawned a dialogue between science and religion. In 2015 Pope Francis proposed an alliance between reason and faith. In a ninety-eight-page encyclical, he made a plea to stop climate destruction. And in an article in *Nature* in 2016 titled "Religion and Science Can Have a True Dialogue," Kathryn Pritchard reported on a "Take Your Vicar to the Lab" project led by parishes of the Church of England.[10]

My hope is that many of these visits will include trips to microbiology laboratories. There the dialogue might start by referencing Antonie van Leeuwenhoek, the Dutch Reformed Calvinist who first saw microbes, and who believed that his amazing discovery of germs was merely proof of the great wonder of God's creation.

APPENDIX

Landmark Discoveries and Notable Germs

LANDMARK DISCOVERIES

1674	Antonie van Leeuwenhoek, father of microbiology; coinventor of the microscope; first to see single-celled life (microbes)
1796	Edward Jenner, father of vaccinology; created the smallpox vaccine
1847	Ignaz Semmelweis, father of hygiene; pioneered antiseptic procedures (hand washing) to prevent puerperal ("childbed") fever
1854	John Snow, father of epidemiology; traced cholera outbreak in London to contaminated water at the Broad Street pump
1856–1863	Johann Gregor Mendel, father of genetics (the fundamental laws of inheritance)
1857–1885	Louis Pasteur, father of pasteurization and fermentation; major contributor to the germ theory of disease and to vaccination
1859	Charles Darwin, publication of *On the Origin of Species by Natural Selection*; father of the theory of evolution

1876–1882   Robert Koch, father of bacteriology and of the germ
             theory of disease; devised methods for culturing bacteria;
             discovered the cause of anthrax and tuberculosis

1882–1909   Elie Metchnikoff, father of cellular immunology and
             probiotics

1892–1898   Dimitri Ivanovsky and Martinus Beijernick, founders of
             virology

1915–1918   Frederick Twort and Félix d'Herelle, discoverey of
             bacteriophages

1928   Alexander Fleming, discovery of penicillin

1935   Arthur Tansley, father of ecology; defined ecosystems as
             interacting communities of living organisms (plants,
             animals, and microbes) with nonliving components of
             their environment, such as, air, water, and mineral soil

1953   James Watson and Frederick Crick, discovery of the
             double-helix structure of DNA (the blueprint for protein
             production and heredity)

1977–1991   Carl Woese, pioneer of field of metagenomics (the study
             of genetic material directly from environmental samples
             using molecular biology techniques) and discoverer of
             the domain of life called Archaea

2016   William Martin and colleagues, postulation of LUCA
             (last universal common ancestor)

NOTABLE GERMS AND GERM BEHAVIOR (THE GOOD, THE BAD, AND THE INDIFFERENT)

- Microbe–host relationships: mutualism (the good—both benefit); parasitism (the bad—the microbe benefits and the host is harmed); commensalism (the indifferent—the microbe benefits and the host neither gains nor loses); symbiosis (any type of persistent biological interaction between two or more different species); endosymbiosis (symbiosis in which one of the symbiotic organisms lives inside the other).

- More than 99 percent of microbes can't be cultured (they're detected by metagenomic techniques).
- Bacteria (the first life-form) appeared on Earth 3.8 billion years ago.
- Archaea and Bacteria (prokaryotes)—ancient common ancestors of all life; gave rise to eukaryotes (protists, fungi, animals, and plants).
- Viruses (mainly bacteriophages)—emerged and coevolved with bacteria and archaea; number 10^{31}.
- Stromatolites (fossilized cyanobacteria) formed in Greenland, 3.7 billion years ago—Earth's oldest fossils.
- Our ancient microbial ancestors were extremophiles that lived and reproduced in hostile environments, such as extreme heat, cold, acidity, and salinity.
- Cyanobacteria were responsible for the Great Oxygenation Event, 2.3 billion years ago, allowing for the emergence of aerobes; today microbes in the sea produce about 50 percent of the oxygen we breathe and remove roughly the same proportion of CO_2 from the atmosphere.
- Microbes had our planet to themselves for two billion years. Different species (estimates): all prokaryotes and eukaryotes (total 10–50+ million); bacteria (10 million, with only 1,400 human pathogens); archaea (unknown number with very, very few being human pathogens); fungi (1.5 to 5 million and only 300 human pathogens); and viruses (1 to 100 million, with only 128 human pathogens).
- In 2001, the human microbiome was defined by Joshua Lederberg as the ecological community of commensal, symbiotic, and pathogenic microorganisms that literally share our body space.
- The Human Microbiome Project (launched in 2008, and ongoing): gut (39 trillion bacteria—about the same number as the number of cells in the human body, about 2,000 species, more than 100 fungal species, trillions of virus species, and an unknown number of archaea); skin (1 trillion bacteria, about 1,000 species, and hundreds of species of fungi); vagina (80 species of *Lactobacillus* alone); mouth (more than 1,000 bacterial species); and lungs (undetermined numbers and species of bacteria, archaea, fungi, and viruses).
- Dysbiosis (imbalance of the microbial composition of the gut microbiome), associated with obesity, type 2 diabetes, inflammatory bowel diseases (Crohn's disease and ulcerative colitis), irritable bowel syndrome, cardiovascular disease, colon cancer, asthma, allergies, and autoimmune diseases, such as multiple sclerosis and lupus.

- Notorious mortal enemies: smallpox (variola major virus, which killed more people than all wars in history combined and was eradicated in 1977); the plague (*Yersinia pestis*, with twenty-eight epidemics, and the Black Death [1346–1353] killed 30 to 60 percent of all Europeans); tuberculosis (*Mycobacterium tuberculosis*, the nineteenth-century Great White Plague, which killed one-third of all Europeans and continues to kill 1.5 million people a year); malaria (*Plasmodium falciparum*, which continues to kill almost half a million people [mostly children] a year); and cholera (*Vibrio cholera*, with seven pandemics in the nineteenth and twentieth centuries, which continues to kill more than one hundred thousand people a year).
- Emerging infections: infections appearing, reappearing, or changing geography since 1967: estimated at 140–168 microbial species; 60 percent are zoonotic—that is, they are passed from animals to humans, either directly or indirectly, via insects.

EMERGING INFECTIONS HIGHLIGHTED IN THIS BOOK

In the order of appearance in part 2.

- HIV: discovered in 1983 by Luc Montagnier and Francoise Barre-Sinoussi; killed 39 million people by 2013
- Ebola virus: discovered in Zaire and Sudan in 1976 by CDC scientists and Peter Piot; the West Africa epidemic of 2013–2015 killed over 11,000 people
- Dengue virus: discovered in 1907 by P. M. Ashburn and Charles F. Craig; by 2010, 50–100 million people infected worldwide per year (with 500,000 life-threatening cases)
- Chikungunya virus: discovered in 1952 in Tanganyika/Tanzania; arrived in the Caribbean in 2013 and spread throughout Latin America, leading to over 1,000,000 cases there in its first year
- Zika virus: discovered in 1947 in the Zika forest of Uganda; arrived in Brazil in 2015, leading to over 1,000,000 cases in its first year; now rapidly spreading throughout Americas and Southeast Asia.
- West Nile virus: discovered in 1937 in Uganda; arrived in New York City in 1999, and spread to all lower forty-eight states

- Avian/bird flu in humans: A(H1N1): 1918 pandemic infected 500 million people across the world, killing 50 to 100 million (one of the deadliest natural disasters in human history); A(H5N1): first case in 1997, more than 700 cases since 2003, deadly in over half of all cases; A(H7N9): 139 cases in China in 2013; 735 human cases by June 2017; deadly in over a third of cases
- Nipah virus: cause of encephalitis; discovered in 1999 in Malaysia; over 500 cases in southeast Asia; mortality of 40 to 70 percent
- SARS-CoV: cause of severe acute respiratory syndrome; 2003 outbreak arose in Asia; 8,098 cases worldwide; mortality of 10 percent; no cases since 2004
- MERS-CoV: cause of Middle East respiratory syndrome; discovered in 2012 in Saudi Arabia by Ali Mohamed Zaki; by 2016, 1,644 cases reported in twenty-six countries, with 590 deaths
- *Legionella pneumophila*: cause of Legionnaires' disease; first appeared in Philadelphia in 1976; 8,000 to 18,800 cases per year continue to appear in the United States
- *Borrelia burgdorferi*: cause of Lyme disease; discovered in 1981 by Willy Burgdorfer; 30,000 to 300,000 cases per year in the United States
- Anaplasmosis: human cases discovered in 1980 by J. S. Dumler and Johan Bakken; over 2,000 cases per year in the United States
- Babesiosis: first human case in 1957; 1969 U.S. epidemic began on the northeastern coast; spread to the Midwest (1,762 cases reported to the CDC in 2013)
- *Escherichia coli* 0157:H7: cause of bloody diarrhea; first recognized 1982; over 95,000 cases per year in the United States
- Mad cow disease (bovine spongiform encephalopathy): appeared in the UK in 1986; first reported in humans in 1996, as a variant of Creutzfeldt–Jakob disease; uniformly fatal; 229 cases by 2014
- *Clostridium difficile* colitis: identified in humans in 1978 by John Bartlett and others; 450,000 cases and 15,000 deaths per year in the United States; fecal microbiota transplants demonstrated as a highly effective treatment by 2013
- *Cryptosporidium parvum*: cause of cryptosporidiosis; first human case in 1976; causes 748,000 cases of gastroenteritis per year in the United States

- Norovirus: discovered in 1972 by Albert Z. Kapikian; causes 19 to 21 million cases of acute gastroenteritis per year in the United States
- Methicillin-resistant *Staphylococcus aureus* (MRSA): emerged in 1968 in U.S. hospitals; emerged in communities in 1997; in 2005, there were over 278,000 MRSA-related hospitalizations in the United States
- Carbapenem-resistant *Enterobacteriacae* (CRE): emerged in 2001, considered "nightmare bacteria" because of their resistance to most antibiotics
- Colistin-resistant *Escherichia coli*: emerged in 2015, carries genes that can be passed on to other gram-negative species, potentially resulting untreatable bacterial infections

NOTES

INTRODUCTION

1. Given the apparent enthusiasm of the kids, I was quite proud of this talk. So I was surprised that evening when my daughter failed to even mention it. When I commented about all the questions her classmates had, she replied, "It was great, Dad, because you got us out of a half hour of the next class."

2. Joshua Lederberg, Robert E. Shope, and Stanley C. Oaks Jr., eds., *Emerging Infections: Microbial Threats to Health in the United States* (Washington, DC: National Academies Press, 1992).

I. THE TREE OF LIFE

1. Roland R. Griffiths et al., "Psilocybin Produces Substantial and Sustained Decreases in Depression and Anxiety in Patients with Life-Threatening Cancer: A Randomized Double-Blind Trial," *Journal of Psychopharmacology* 30, no. 12 (December 2016): 1181–97.

2. David Quammen, *The Tangled Tree: A Radical New History of Life* (New York: Simon & Schuster, 2018).

3. Every spoonful of seawater harbors millions of viruses. Recent studies suggest that the world's oceans contain nearly two hundred thousand virus species, with an unexpected pocket of viral diversity in the Arctic Ocean. Ann C. Gregory et al., "Marine DNA Viral Macro- and Microdiversity from Pole to Pole," *Cell* 177 (May 16, 2019): 1–15.

4. Arshan Nasir and Gustavo Caetano-Anolles, "A Phylogenomic Data-Driven Exploration of Viral Origins and Evolution," *Science Advances* 1, no. 8 (September 25, 2015).

5. Frederik Schulz et al., "Giant Viruses with an Expanded Complement of Translation System Components," *Science* 356, no. 6333 (April 7, 2017): 82–85.

2. IT'S A MICROBIAL WORLD

1. You may recall the anthrax attacks in September 2001. Beginning one week after the September 11 attack on the World Trade Center, spores of the highly lethal bacterium *B. anthracis* were mailed to two U.S. senators and several media offices. Ultimately, five people were killed and another seventeen infected. In 2008, the suspected perpetrator, a scientist who worked at the U.S. government's biodefense labs at Fort Detrick, committed suicide. The anthrax scare triggered an enormous governmental and scientific investment aimed at preventing anthrax as well as other microbial forms of bioterrorism. At the time—along with most infectious disease specialists in America—I had to familiarize myself with the clinical features of anthrax because this disease had virtually disappeared in the developed world.

2. Yinon M. Bar-On, Rob Phillips, and Ron Milo, "The Biomass Distribution on Earth," *Proceedings of the National Academy of Sciences of the USA* 115, no. 25 (June 19, 2018): 6506–11.

3. Deep Carbon Observatory, "Life in Deep Earth Totals 15 to 23 Billion Tons of Carbon—Hundreds of Times More Than Humans," *ScienceDaily*, December 10, 2018, https://www.sciencedaily.com/releases/2018/12/181210101909.htm.

4. Edward O. Wilson, *Genesis: The Deep Origin of Societies* (New York: Liveright, 2019).

5. Antonio Damasio, *The Strange Order of Things: Life, Feeling, and the Making of Cultures* (New York: Pantheon Books, 2018).

6. For more on the view that natural selection can favor collective or social behavior, see Nicholas A. Christakis, *Blueprint: The Evolutionary Origins of a Good Society* (New York: Little, Brown Spark, 2019).

3. THE HUMAN MICROBIOME

1. David Quammen, *The Tangled Tree: A Radical New History of Life* (New York: Simon & Schuster, 2018). A major focus of this book is on the life and

work of Carl Woese, the late University of Illinois microbiologist who is credited with the 1977 discovery of the domain Archaea. The technology that he and his colleagues used to detect these microbes, metagenomics, revolutionized the field of microbiology.

2. Susan L. Prescott, "History of Medicine: Origin of the Term Microbiome and Why It Matters," *Human Microbiome Journal* 4 (June 2017): 24–25. Joshua Lederberg is almost universally credited with coining the term in 2001. However, the word actually dates back to at least 1988.

3. Michael Specter, "Germs Are Us," *New Yorker*, October 15, 2012.

4. Martin J. Blaser, *Missing Microbes: How the Overuse of Antibiotics Is Fueling Our Modern Plagues* (New York: Henry Holt, 2014). An internist and infectious diseases expert, Martin Blaser is arguably the foremost physician scientist working in the field of human microbiome research. Many of his seminal studies were carried out at New York University's Lagone Medical Center, where he was director of the Human Microbiome Research Program.

5. Cassandra Willyard, "Could Baby's First Bacteria Take Root Before Birth?," *Nature* 553 (January 17, 2018): 264–66. Conventional wisdom has suggested that the womb is sterile and that newborns are first colonized by microbes in the birth canal (or skin, in the case of cesarean births). But, as discussed in this paper, controversial new studies suggest that the fetus may encounter harmless or beneficial microbes within the placenta before birth.

6. Simon Lax et al., "Longitudinal Analysis of Microbial Interaction between Humans and the Indoor Environment," *Science* 345, no. 6200 (August 29, 2014): 1048–52.

7. Jack A. Gilbert et al., "Current Understanding of the Human Microbiome," *Nature Medicine* 24 (April 10, 2018): 392–400.

8. Lisa Maier et al., "Extensive Impact of Non-Antibiotic Drugs on Human Gut Bacteria," *Nature* 555 (March 29, 2018): 623–28.

9. Matt Richtel, *An Elegant Defense: The Extraordinary New Science of the Immune System* (New York: Morrow, 2019). Although the hygiene hypothesis has been around for several decades, the idea that cleanliness is not always next to godliness is now being popularized—probably because studies of the human microbiome are reaching the attention of the general public. But the recommendations of some doctors not to worry about children eating dirt, or the products of their nose picking, await scientific confirmation.

10. Michelle M. Stein et al., "Innate Immunity and Asthma Risk in Amish and Hutterite Farm Children," *New England Journal of Medicine* 375 (August 4, 2016): 411–21.

11. Hein M. Tun et al., "Exposure to Household Furry Pets Influences the Gut Microbiota of Infants at 3–4 Months Following Various Birth Scenarios," *Microbiome* 5 (April 6, 2017): 40.

12. Bas E. Dutilh et al., "A Highly Abundant Bacteriophage Discovered in the Unknown Sequences of Human Faecal Metagenomes," *Nature Communications* 5, no. 4498 (July 24, 2014).

13. Robynne Chutkan, *The Microbiome Solution: A Radical New Way to Heal Your Body from the Inside Out* (New York: Avery, 2015).

14. Fanil Kong et al., "A New Study of Chinese Long-Lived Individuals Identifies Gut Microbial Signatures of Healthy Aging," *Current Biology* 26, no. 18 (September 26, 2016): R832-R833.

15. Vanessa K. Ridaura et al., "Gut Microbiota from Twins Discordant for Obesity Modulate Metabolism in Mice," *Science* 341, no. 6150 (September 6, 2013).

16. R. Liu et al., "Gut Microbiome and Serum Metabolome Alterations in Obesity and after Weight-Loss Intervention," *Nature Medicine* 23, no. 7 (June 19, 2017): 859–68.

17. Bertrand Routy et al., "Gut Microbiome Influences Efficacy of PD-1-based Immunotherapy against Epithelial Tumors," *Science* 359, no. 6371 (January 5, 2018): 91–97.

18. Emma Barnard et al., "The Balance of Metageomic Elements Shapes the Skin Microbiome in Acne and Health," *Scientific Reports* 6, no. 39491 (December 21, 2016), DOI: 10.1038/srep39491.

19. Chris Callewaert, Jo Lambert, and Tom Van de Wiele, "Towards a Bacterial Treatment for Armpit Malodour," *Experimental Dermatology* 26, no. 5 (May 2017): 388–91.

20. Yang He et al., "Gut–Lung Axis: The Microbial Contributions and Clinical Implications," *Critical Reviews in Microbiology* 43, no. 1 (October 26, 2016): 81–95.

21. Timothy R. Sampson et al., "Gut Microbiota Regulate Motor Deficits and Neuroinflammation in a Model of Parkinson's Disease," *Cell* 167, no. 6 (December 1, 2016): 1469–80.

22. Emeran Mayer, *The Mind-Gut Connection: How Hidden Conversation within Our Bodies Impacts Our Mood, Our Choices, and Our Health* (New York: HarperCollins, 2016).

23. I. A. Marin et al., "Microbiota Alteration Is Associated with the Development of Stress-Induced Despair Behavior," *Scientific Reports* 7, no. 43859 (March 7, 2017).

24. Elizabeth Pennisi, "Gut Bacteria Linked to Mental Well-Being and Depression," *Science* 363, no. 6427 (February 8, 2019): 569.

25. Susan L. Lynch and Oluf Pedersen. "The Human Intestinal Microbiome in Health and Disease," *New England Journal of Medicine* 375 (December 15, 2016): 2369–79.

26. Rodney Dietert, *The Human Superorganism: How the Microbiome Is Revolutionizing the Pursuit of a Healthy Life* (New York: Dutton, 2016).

27. Rob Dunn, *The Wild Life of Our Bodies: Predators, Parasites, and Partners That Shape Who We Are Today* (New York: HarperCollins, 2014).

4. DEPARTMENTS OF BODILY DEFENSE

1. Jan C. Rieckmann et al., "Social Network Architecture of Human Immune Cells Unveiled by Quantitative Proteomics," *Nature Immunology* 18, no. 5 (May 1, 2017): 583–93.

2. Ian F. Miller and C. Jessica E. Metcalf, "Evolving Resistance to Pathogens," *Science* 363, no. 6433 (March 22, 2019): 1277–78.

3. Leore T. Geller et al., "Potential Role of Intratumor Bacteria in Mediating Tumor Resistance to the Chemotherapeutic Drug Gemcitabine," *Science* 357, no. 6356 (September 15, 2017): 1156–60.

5. IT'S ALL CONNECTED

1. Delphine Destoumieux-Garzon et al., "The One Health Concept: 10 Years Old and a Long Road Ahead," *Frontiers in Veterinary Science* 5, no. 14 (February 12, 2018), DOI: 10.33891/fvets2018.00014.

2. Jop de Vrieze, "This Project Brings Desert Soil to Life," OperationCO2.com, June 30, 2015. http://operationco2.com/life-news/this-project-brings-desert-soil-to-life-418.html.

3. Emily Monosson, *Natural Defense: Enlisting Bugs and Germs to Protect Our Food and Health* (Washington, DC: Island Press, 2017).

4. Kasie Raymann, Zack Shaffer, and Nancy A. Moran, "Antibiotic Exposure Perturbs the Gut Microbiota and Elevates Mortality in Honeybees," *PLOS Biology* 15, no. 3 (March 14, 2017), DOI:10.13711/journal.pbio.2001861.

5. Habib Yaribeygi et al., "The Impact of Stress on Body Function: A Review," *EXCLI Journal* 16 (July 21, 2017): 1057–72, DOI: 10.17179/excli2017-480.

6. WHAT HAS PLAGUED US?

1. Joshua Lederberg, Robert E. Shope, and Stanley C. Oaks Jr., eds., *Emerging Infections: Microbial Threats to Health in the United States* (Washington, DC: National Academies Press, 1992).

2. Arthur W. Boylston, "The Myth of the Milkmaid," *New England Journal of Medicine*, no. 378 (February 1, 2018): 414–15.

3. Livia Schrick et al., "An Early American Smallpox Vaccine Based on Horsepox," *New England Journal of Medicine* 377 (October 12, 2017): 1491–92.

4. When I served as a medical officer at the American Indian Hospital in Santa Fe in the early 1970s, we doctors were always on the lookout for plague. I remember vividly the report of the case of a young man from the San Felipe Pueblo, where I worked in the clinic. The young man had been admitted to a hospital in Albuquerque with a fever and swollen lymph glands. He was deathly ill, but because a diagnosis and treatment weren't established after two days, his family took him back to the pueblo. When his blood culture was later reported to contain *Y. pestis*, a public health nurse was sent to the pueblo to immediately initiate antibiotic therapy. But the young man was nowhere to be found; he was being treated by the medicine man. When he showed up two weeks later at the same hospital in Albuquerque for antibiotic treatment, it was no longer needed. He had completely recovered, and no one was able to find out what the medicine man had done for him.

5. Michael J. A. Reid et al., "Building a Tuberculosis-Free World: The *Lancet* Commission on Tuberculosis," *Lancet* 393 (March 30, 2019): 1331–84.

6. Martin J. Blaser, "The Theory of Disappearing Microbiota and the Epidemics of Chronic Diseases," *Nature Reviews Immunology* 17 (July 27, 2017): 461–63.

7. KILLER VIRUSES

1. To give you some idea of how sneaky HIV is, it took until 2019, almost four decades into the HIV/AIDS epidemic, for the second person to be cured of HIV infection. This case, like the first cured patient, required a bone marrow transplant containing CD4 lymphocytes that resist infection. Jon Cohen, "Has a Second Person with HIV Been Cured?," *Science* 363, no. 6431 (March 8, 2019): 1021.

2. Susan Jaffe, "USA Sets Goal to End the HIV Epidemic in a Decade," *Lancet* 393 (February 16, 2019): 625–26.

3. Jon Cohen, "AIDS Pioneer Finally Brings AIDS Vaccine to Clinic," *Science/Health,* October 8, 2015, https://www.sciencemag.org/news/2015/10/aids-pioneer-finally-brings-aids-vaccine-clinic.

4. To its credit, Merck is providing the vaccine for free. The vaccine is made by adding a gene for an Ebola virus surface protein to vesicular stomatitis virus, which is harmless to humans. Jon Cohen, "Ebola Outbreak Continues Despite Powerful Vaccine," *Science* 364, no. 6437 (April 19, 2019): 223.

5. Vinh-Kim Nguyen, "An Epidemic of Suspicion—Ebola and Violence in the DRC," *New England Journal of Medicine* 380, no. 14 (April 4, 2019): 1298–99.

8. THE BUZZ ON MOSQUITO-BORNE INFECTIONS

1. Recent research at Rockefeller University suggests that a receptor in the mosquito's antennae, named IR8a, allows detection of lactic acid, another odorant mosquitoes find attractive. Joshua I. Raji et al., "*Aedes aegypti* Mosquitoes Detect Acidic Volatiles Found in Human Odor Using the IR8a Pathway," *Current Biology* 29, no. 8 (April 22, 2019): 1253–62. Such studies will likely lead to improved mosquito repellents.

2. The dengue virus is peculiar because a first infection is rarely fatal, but when a second infection with a different dengue virus type occurs, it can lead to a much more serious disease. And the vaccine Dengvaxia promotes immunity to each of the four virus types. Thus, in people who have never been infected with a dengue virus, the vaccine can act like a first infection, potentially priming such individuals for a serious and even fatal response to a second natural infection. Nowadays, before the vaccine is administered, a test is done to make sure the person receiving the vaccine has had a previous natural dengue infection. Sadly, a prominent pediatrician and dengue researcher in the Philippines, where the vaccine was yanked in 2017, was indicted for her role in promoting the vaccine. Fatima Arkin, "Dengue Researcher Faces Charges in Vaccine Fiasco," *Science* 364, no. 6438 (April 26, 2019): 320.

3. Emil C. Reisinger et al., "Immunogenicity, Safety, and Tolerability of the Measles-Vectored Chikungunya Virus Vaccine MV-CHIK: A Double-Blind, Randomised, Placebo-Controlled and Active-Controlled Phase 2 Trial," *Lancet* 392 (December 22/29, 2018): 2718–27.

4. Guillain-Barré syndrome (GBS) is a neurological disorder in which the immune system attacks the nerves, resulting in progressive and potentially life-threatening weakness or paralysis. By September 2016, twelve countries had reported an increase in the number of cases of GBS due to Zika.

9. MICROBES IN FLIGHT

1. During the course of my clinical research career, I studied infections that impact patients with compromised immune systems, such as those taking immunosuppressive drugs to prevent the rejection of transplanted organs. At the beginning of my studies in 1977, there were, of course, no patients with West Nile virus disease. But by the early years of this century, West Nile virus had become the most common cause of brain infection in organ transplant recipients. Most of the patients I consulted on were in their seventies, except for organ transplant recipients, who were younger.

2. David C. E. Philpott et al., "Acute and Delayed Deaths after West Nile Virus Infection, Texas, USA, 2002–2012," *Emerging Infectious Diseases* 25, no. 2 (February 2019): 256–64.

3. Galia Rahav et al., "Primary versus Nonprimary West Nile Virus Infection: A Cohort Study," *Journal of Infectious Diseases* 213, no. 5 (March 1, 2016): 755–61.

4. Wenqing Zhang and Robert G. Webster, "Can We Beat Influenza?," *Science* 357, no. 6347 (July 14, 2017): 111.

5. Bats do get sick, of course. Besides being susceptible to illnesses such as rabies, bats have their own unique emerging infection called white-nose syndrome, which cannot be passed on to humans. This is a highly lethal infection caused by a fungus with a scientific name that is almost as ugly as the disease: *Pseudogymnoascus destructans*. White-nose syndrome gets its name from a distinctive white fuzz that appears on the face of infected bats. First recognized in New York State in 2007, white-nose syndrome has been on a relentless westward march. By 2016, it had reached the state of Washington and killed more than six million bats. And in early 2019, many Minnesota bat species were on the brink of extinction.

10. DON'T BREATHE THIS AIR

1. Yaseen M. Arabi et al., "Middle East Respiratory Syndrome," *New England Journal of Medicine* 376 (February 9, 2017): 584–94.

2. Nick Phin et al. "Epidemiology and Clinical Management of Legionnaires' Disease," *Lancet* 14, no. 10 (June 23, 2014): 1011–21.

11. MICROBES IN THE WOODS

1. While *Ixodes scapularis* is human enemy number one among the various types of ticks, a variety of other tick species also afflict a good deal of human suffering by transmitting a variety of pathogens. One worrisome new arrival to the United States, the Asian longhorned tick (*Haemaphysalis longicornis*), was discovered in 2017 and already has spread to at least nine states. How it made its way to our shores is unknown. Because it can transmit many types of pathogens common in the United States, it is clearly on the radar screen of the Centers for Disease Control and Prevention. (In 2019, for the first time the CDC began monitoring the nation's tick population for the pathogens they carry that may afflict humans.) Another relative newcomer is the tick species *Ornithodoros turicata*. It recently was responsible for an outbreak in Austin, Texas, of tick-borne relapsing fever, caused by a *Borrelia* species, *B. turicatae*. Jack D. Bissett et al., "Detection of Tickborne Relapsing Fever Spirochete, Austin, Texas, USA," *Emerging Infectious Diseases* 24, no. 11 (2018): 2003–9.

2. Posttreatment Lyme disease syndrome is not rare. Although the number of patients with Lyme disease who fail to recover fully after appropriate antibiotic therapy is unknown, a recent estimate suggests it could hit two million by 2020. Allison DeLong, Mayla Hsu, and Harriet Kotsoris, "Estimation of Cumulative Number of Post-Treatment Lyme Disease 2020," *BMC Public Health* 19 (April 24, 2019): 352.

3. In recent years, a little-understood entity called chronic Lyme disease has aroused much concern and controversy. While patients with this ailment are clearly ill, they typically don't test positive for *B. burgdorferi* infection, or their symptoms don't meet the formal clinical criteria for Lyme disease. Part of the problem, however, is that lab tests often aren't that helpful in establishing a diagnosis of Lyme disease. Generally, infectious disease specialists advise against treating chronic Lyme disease, in the belief that treatment is ineffective and sometimes very harmful. In fact, randomized trials of longer-term antibiotic therapy for patients with chronic Lyme disease showed no additional benefit to their health-related quality of life beyond those treated for a shorter period. Anneleen Berende et al., "Randomized Trial of Longer-Term Therapy for Symptoms Attributed to Lyme Disease," *New England Journal of Medicine* 274 (March 31, 2016): 1209–20. The counterargument, however, is compelling: Does it really make sense to shrug and do nothing? Why not attempt to treat an illness that someone suffers from? But because antibiotics can be dangerous, most physicians, myself included, recommend against this approach.

4. J. Stephen Dumler et al., "Human Granulocytic Anaplasmosis and *Anaplasma phagocytophilum*," *Emerging Infectious Diseases* 11, no. 12 (December 2005): 1828–34.

5. Edouard Vannier and Peter J. Krause, "Human Babesiosis," *New England Journal of Medicine* 366 (June 21, 2012): 2397–2407.

12. WHAT'S IN THE BEEF?

1. Joan Stephenson, "Nobel Prize to Stanley Prusiner for Prion Discovery," *Journal of the American Medical Association* 278, no. 18 (November 12, 1997): 1479.

2. Richard T. Johnson, "Prion Diseases," *Lancet Neurology* 4, no. 10 (October 1, 2005): 635–42.

3. Ross M. S. Lowe et al., "*Escherichia coli* O157:H7 Strain Origin, Lineage, and Shiga Toxin 2 Expression Affect Colonization of Cattle," *Applied and Environmental Microbiology* 75, no. 15 (August 2009): 5074–81.

4. April K. Bogard et al., "Ground Beef Handling and Cooking Practices in Restaurants in Eight States," *Journal of Food Protection* 76, no. 12 (2013): 2132–40.

13. GUT REACTIONS

1. Herbert L. DuPont, "Acute Infectious Diarrhea in Immunocompetent Adults," *New England Journal of Medicine* 370 (April 17, 2014): 1532–40.

2. The genus name of *Clostridium difficile* was changed in 2016 to *Clostridioides difficile* based on its genetic similarity to other members of the *Clostridioides* species of bacteria. Clinically, it is often referred to simply as *C. diff.*

3. In a study of cryptosporidiosis in Peruvian children, 63 percent of infected kids showed no symptoms. Antibodies to *Cryptosporidium* have been detected in about 30 percent of infected children and adults in the United States who show no symptoms.

4. Elizabeth Robilotti, Stan Deresinski, and Benjamin A. Pinsky, "Norovirus," *Clinical Microbiology Reviews* 28, no. 1 (January 2015): 134–64.

5. Jae Hyun Shin et al., "Innate Immune Response and Outcome of *Clostridium difficile* Infection Are Dependent on Fecal Bacterial Composition in the Aged Host," *Journal of Infectious Diseases* 217, no. 2 (January 15, 2018): 188–97.

6. Robert A. Britton and Vincent B. Young, "Interaction between the Intestinal Microbiota and Host in *Clostridium difficile* Colonization Resistance," *Trends in Microbiology* 20, no. 7 (July 2012): 313–19.

14. WHEN BEAUTY ISN'T SKIN DEEP

1. Andie S. Lee et al., "Methicillin-Resistant *Staphylococcus aureus*," *Nature Reviews Disease Primers* 4 (May 31, 2018): article 18033.

2. Carl Andreas Grontvedt et al., "Methicillin-Resistant *Staphylococcus aureaus* CC398 in Humans and Pigs in Norway: A 'One Health' Perspective on Introduction and Transmission," *Clinical Infectious Diseases* 63, no. 11 (December 1, 2016): 1431–38.

3. Bob C. Y. Chan and Paul Maurice, "Staphylococcal Toxic Shock Syndrome," *New England Journal of Medicine* 369 (August 29, 2013): 852.

4. A. P. Kourtis et al., "Vital Signs: Epidemiology and Recent Trends in Methicillin-Resistant and in Methicillin-Susceptible *Staphlococcus aureaus* Bloodstream Infections—United States," *MMWR Morbidity and Mortality Weekly Report* 68, no. 9 (March 8, 2019): 214–19.

15. THE PERILS OF ANTIBIOTIC MISUSE

1. World Health Organization, "High Levels of Antibiotic Resistance Found Worldwide, New Data Show," news release, January 29, 2018.

2. Ruobing Wang et al., "The Global Distribution and Spread of the Mobilized Colistin Resistance Gene mcr-1," *Nature Communications* 9 (March 21, 2018): article 1179.

3. Bradley M. Hover et al., "Culture-Independent Discovery of the Malacidins as Calcium-Dependent Antibiotics with Activity against Multidrug-Resistant Gram-Positive Pathogens," *Nature Microbiology* 3 (February 12, 2018): 415–22.

4. Joseph Nesme et al., "Large-Scale Metagenomic-Based Study of Antibiotic Resistance in the Environment," *Current Biology* 24 (May 19, 2014): 1096–1100.

5. D. J. Livorsi et al., "A Systematic Review of the Epidemiology of Carbapenem-Resistant Enterobacteriaceae in the United States," *Antimicrobial Resistance & Infection Control* 7, no. 55 (April 24, 2018).

6. M. P. Freire et al., "Bloodstream Infection Caused by Extensively Drug-Resistant *Acinetobacer baumannii* in Cancer Patients: High Mortality Associated with Delayed Treatment Rather Than with the Degree of Neutropenia," *Clinical Microbiology and Infection* 22, no. 4 (April 2016): 352–58.

7. You may wonder why scientists eight thousand miles apart are closely examining seagull butts. The answer is that antibiotic-resistant microbes are everywhere. There is almost no place or object (or butt) on Earth that is not teeming with germs—and, thus, teeming with potential for the development or

spread of antibiotic-resistant bacteria. And it is important to determine how these microbes are transmitted around the world.

8. Lancet Commission, "Building a Tuberculosis-Free World: *The Lancet Commission on Tuberculosis*," *Lancet* 393, no. 10178 (March 20, 2019): 1331–84.

9. David W. Eyre et al., "A *Candida auris* Outbreak and Its Control in an Intensive Care Setting," *New England Journal of Medicine* 379 (October 4, 2018): 1322–31.

10. Jeremy D. Keenan et al., "Azithromycin to Reduce Childhood Mortality in Sub-Saharan Africa," *New England Journal of Medicine* 378 (April 26, 2018): 1583–92.

16. THE STRAIGHT POOP ON FECAL TRANSPLANTS

1. Els van Nood et al., "Duodenal Infusion of Donor Feces for Recurrent *Clostridium difficile*," *New England Journal of Medicine* 368 (January 13, 2013): 407–15.

2. Sahil Khanna et al., "A Novel Microbiome Therapeutic Increases Gut Microbial Diversity and Prevents Recurrent *Clostridium difficile* Infection," *Journal of Infectious Diseases* 214, no. 2 (July 15, 2016): 173–81.

3. Stuart Johnson and Dale N. Gerding, "Fecal Fixation: Fecal Microbiota Transplantation for *Clostridium difficle* Infection," *Clinical Infectious Diseases* 64, no. 3 (February 1, 2017): 272–74.

4. Ed Yong, "Sham Poo Washes Out," *Atlantic*, August 1, 2016.

5. Dae-Wook Kang et al., "Long-Term Benefit of Microbiota Transfer Therapy in Autism Symptoms and Gut Microbiota," *Scientific Reports* 9 (April 9, 2010): article 5821, DOI:10.1038/s41598-019-42183-0.

6. Jocelyn Kaiser, "Fecal Transplants Could Help Patients on Cancer Immunotherapy Drugs," *Science/Health*, April 5, 2019, DOI:10.1126/science.aax5960. At the University of Minnesota Cancer Center, Timothy Starr, an assistant professor in the Department of Obstetrics and Gynecology and Women's Health, has teamed up with Alexander Khoruts to study this link more deeply. And at the University of Minnesota Masonic Children's Hospital, Lucie Turcotte, a pediatric hematologist-oncologist, and Khoruts are searching for ways to improve outcomes for leukemia patients by applying knowledge of the microbiome.

17. HEALING WITH FRIENDLY BACTERIA AND FUNGI

1. Kate Costeloe et al., "*Bifidobacterium breve* BBG-001 in Very Preterm Infants: A Randomised Controlled Phase 3 Trial," *Lancet* 387, no. 10019 (February 13, 2016): 649.

2. Stephen B. Freedman et al., "Multicenter Trial of a Combination Probiotic for Children with Gastroenteritis," *New England Journal of Medicine* 379 (November 22, 2018): 2015–26.

3. Jennifer Abbasi, "Are Probiotics Money Down the Toilet? Or Worse?" *Journal of the American Medical Association* 321, no. 7 (February 19, 2019): 633–35.

4. Pinaki Panigrahi et al., "A Randomized Synbiotic Trial to Prevent Sepsis among Infants in Rural India," *Nature* 548 (August 24, 2017): 407–12.

5. Scott C. Anderson, John F. Cryan, and Ted Dinan, *The Psychobiotic Revolution: Mood, Food, and the New Science of the Gut-Brain Connection* (Washington, DC: National Geographic, 2017).

6. Nadja B. Kristensen et al., "Alterations in Fecal Microbiota Composition by Probiotic Supplementation in Healthy Adults: A Systematic Review of Randomized Controlled Trials," *Genome Medicine* 8, no. 1 (May 10, 2016): 52.

7. Andrew I. Geller et al., "Emergency Department Visits for Adverse Events Related to Dietary Supplements," *New England Journal of Medicine* 373 (October 15, 2015): 1531–40.

8. Aida Bafeta et al., "Harms Reporting in Randomized Controlled Trials of Interventions Aimed at Modifying Microbiota: A Systemic Review," *Annals of Internal Medicine* 169, no. 4 (August 21, 2018): 240–47.

18. HEALING WITH FRIENDLY VIRUSES

1. Carl Zimmer, *A Planet of Viruses* (Chicago: University of Chicago Press, 2011).

2. Patrick Jault et al., "Efficacy and Tolerability of a Cocktail of Bacteriophages to Treat Burn Wounds Infected by *Pseudomonas aeruginosa* (PhagoBurn): A Randomised Controlled Double-Blind Phase 1/2 Trial," *Lancet Infectious Diseases* 19, no. 1 (January 1, 2019): 35–45.

3. Robert T. Schooley et al., "Development and Use of Personalized Bacteriophage-Based Therapeutic Cocktails to Treat a Patient with a Disseminated Resistant *Acinetobacter baumannii* Infection," *Antimicrobial Agents and Chemotherapy* 61, no. 10 (October 2017), DOI:10.1128/AAC.00954-17.

4. Rebekah M. Dedrick et al., "Engineered Bacteriophages for Treatment of a Patient with a Disseminated Drug-Resistant *Mycobacterium abscessus*," *Nature Medicine* 25 (May 8, 2019): 730–33.

5. Waqas Nasir Chaudhry et al., "Synergy and Order Effects of Antibiotics and Phages in Killing *Pseudomonas aeruginosa* Biofilms," *PLOS One* (January 11, 2017), DOI:10.1371/journal.pone.0168615.ecollection2017.

19. THE FUTURE OF VACCINES

1. John Rhodes, *The End of Plagues: The Global Battle against Infectious Disease* (New York: Palgrave Macmillan, 2013).

2. Pasteur's landmark discovery is remembered in a mosaic in the chapel of the institute in Paris named in his honor. Pasteur found, rather serendipitously, that the inoculation of chickens with an old culture of the bacteria that cause chicken cholera protected them against a subsequent challenge with fresh, virulent cholera bacteria. He realized that the old culture contained bacteria that had become weakened and helped the chickens develop immunity. Of course, he didn't know the intricacies of the immune system that are summarized in chapter 4. But this experiment demonstrated that immunity is all about developing memory of germs, and it launched the field of immunology.

3. Michael R. Weigand et al., "Genomic Survey of *Bordetella pertussis* Diversity, United States, 2000–2013," *Emerging Infectious Diseases* 25, no. 4 (April 2019): 780–83.

4. A mumps outbreak at Temple University in Philadelphia began in February 2019 and spread to over one thousand students by the end of March. Jeremy Bauer-Wolf, "Measles Outbreak at Temple University," *Inside Higher Ed*, April 2, 2019, https://www.insidehighered.com/news/2019/04/02/temple-sees-mumps-outbreak-more-100-cases. The outbreak was believed to have originated with a person who traveled internationally, underscoring the need to be properly vaccinated before foreign travel. (In some countries, such as Japan, people aren't routinely vaccinated against measles.) Close living quarters on most college campuses are setups for spreading infection.

5. Mark Honigsbaum, "Vaccination: A Vexatious History," *Lancet* 387, no. 10032 (May 14, 2016): 1988–89.

6. Shawn Otto, *War on Science: Who's Waging It, Why It Matters, What Can We Do about It* (Minneapolis: Milkweed Editions, 2016).

7. H. Cody Meissner, Narayan Nair, and Stanley A. Plotkin, "The National Vaccine Injury Compensation Program: Striking a Balance between Individual Rights and Community Benefit," *Journal of the American Medical Association* 321, no. 4 (January 29, 2019): 343–44.

8. Catharine I. Paules, Hilary D. Marston, and Anthony S. Fauci, "Measles in 2019—Going Backward," *New England Journal of Medicine* (April 17, 2019), DOI:10.1056/NEJMp1905099. (Epub ahead of print.)

9. Larry O. Gostin, "Law, Ethics, and Public Health in the Vaccination Debates: Politics of the Measles Outbreak," *Journal of the American Medical Association* 313, no. 11 (March 17, 2015): 1099–1100.

10. Phillip K. Peterson, *Get Inside Your Doctor's Head: 10 Commonsense Rules for Making Better Decisions about Medical Care* (Baltimore, MD: Johns Hopkins University Press, 2013).

11. Andrew J. Wakefield et al., "Ileal-Lymphoid-Nodular Hyperplasia, Nonspecific Colitis, and Pervasive Developmental Disorder in Children," *Lancet* 351, no. 9103 (February 28, 1998): 637–41. Retracted in *Lancet* 375, no. 9713 (February 6–12, 2010): 445.

12. Beate Ritz et al., "Air Pollution and Autism in Denmark," *Environmental Epidemiology* 2, no. 4 (December 2018): e028, DOI:10.1097/EE9.0000000000000028.

13. Lindzi Wessell, "Four Vaccine Myths and Where They Came From," *Science*, April 27, 2017, DOI:10.1126/scienceaa/1110.

14. Heidi J. Larson and William S. Schulz, "Reverse Global Vaccine Dissent," *Science* 364, no. 6436 (April 12, 2019): 105.

15. Leslie Roberts, "Is Measles Next?," *Science* 348, no. 6238 (May 29, 2015): 958–63.

16. James Colgrove, "Vaccine Refusal Revisited—the Limits of Public Health Persuasion and Coercion," *New England Journal of Medicine* 375 (October 6, 2016): 1316–17.

20. MICROBES AND THE SIXTH EXTINCTION

1. Naturalist Alfred Russel Wallace, a friendly colleague of Darwin's who corresponded with him, came to essentially the same conclusions in 1858. In fact, the two had planned to unveil the theory together at a scientific conference in July of that year. In the end, though, Wallace had to make the presentation on his own, because on that day Darwin and his wife were burying their young son, who had just died of scarlet fever—a serious infectious disease. (During the early 1840s, a gardener named Patrick Matthew also came up with the principles of natural selection, and even published them—in an appendix to a book called *Naval Timber and Arboriculture*. Unsurprisingly, the world failed to take note of Matthew's insight at the time.)

2. Jean-Jacques Hublin et al., "New Fossils from Jebel Irhoud, Morocco and the Pan-African Origin of *Homo sapiens*," *Nature* 546 (June 8, 2017): 289–92.

3. Anthony D. Barnosky, *Dodging Extinction: Power, Food, Money, and the Future of Life on Earth* (Oakland: University of California Press, 2014).

4. Camilo Mora et al., "How Many Species Are There on Earth and in the Ocean?," *PLOS Biology* (August 23, 2011), DOI:10.1371/journal.pbio.1001127.

5. Kenneth J. Locey and Jay T. Lennon, "Scaling Laws Predict Global Microbial Diversity," *Proceedings of the National Academy of Science USA* 113, no. 21 (May 24, 2016): 5970–75.

6. If, like me, you were schooled on evolution before the 1970s, you were taught that evolution is a very slow process that proceeds over time at even, predictable rates ("Nature does not make leaps"). Thus you may be surprised to know that there have been many shocks to Earth's systems, including the Big Five Mass Extinctions. In 1972, Stephen Gould, one of the most influential evolutionary biologists of the twentieth century, introduced the controversial concept of punctuated equilibrium—the key insight that the great majority of species originate in geological moments (punctuations) and then persist thereafter.

7. Amazingly, seagoing microscopic phytoplankton produce, via photosynthesis, about 50 percent of the world's oxygen. The remainder is produced via photosynthesis on land by plants and trees.

8. Daniel H. Rothman et al., "Methanogenic Burst in the End-Permian Carbon Cycle," *Proceedings of the National Academy of Sciences USA* 111, no. 15 (April 15, 2014): 5462–67.

9. John Kappelmann et al., "Perimortem Fractures in Lucy Suggest Mortality Fall out of a Tall Tree," *Nature* 537 (September 22, 2016): 503–7.

10. A new study of genomes from New Guineans suggests that there was mixing of *Homo sapiens* with the extinct cousins of Neanderthals, the Denisovans, as recently as fifteen thousand years ago. Ann Gibbons, "Moderns Said to Mate with Late-Surviving Denisovans," *Science* 364, no. 6435 (April 5, 2019): 12–13.

11. Yuval Noah Harari, *Sapiens: A Brief History of Humankind* (New York: HarperCollins, 2015).

12. Charlotte J. Houldcroft and Simon J. Underdown, "Neaderthals May Have Been Infected by Diseases Carried out of Africa by Humans," *American Journal of Physical Anthropology* 160, no. 3 (July 2016): 379–88.

13. Eugene Rosenberg and Ilana Zilber-Rosenberg, "Microbes Drive Evolution of Animals and Plants: The Hologenome Concept," *mBio* (March 31, 2016), DOI:10.1128/mBio.01395-15.

14. Jean-Christophe Simon et al., "Host-Microbiota Interactions: From Holobiont Theory to Analysis," *Microbiome* 7, no. 5 (January 11, 2019).

15. David Enard et al., "Viruses Are a Dominant Driver of Protein Adaptation in Mammals," *eLife* (May 17, 2016): e12469, DOI:10.7354/eLife.12469.

16. Elizabeth Kolbert, *The Sixth Extinction: An Unnatural History* (New York: Henry Holt, 2014).

17. Nicole L. Boivin et al., "Ecological Consequences of Human Niche Construction: Examining Long-Term Anthropogenic Shaping of Global Species Distributions," *Proceedings of the National Academy of Sciences USA* 113, no. 23 (June 7, 2016): 6388–96.

18. Ben C. Scheele et al., "Amphibian Fungal Panzootic Causes Catastrophic and Ongoing Loss of Biodiversity," *Science* 363, no. 6434 (March 29, 2019): 1459–62.

19. UN Intergovernmental Science-Policy Platform on Biodiversity and Ecosystems, "UN Report: Nature's Dangerous Decline 'Unprecedented'; Species Extinction Rates 'Accelerating,'" May 6, 2019.

20. Derek R. MacFadden et al., "Antibiotic Resistance Increases with Local Temperature," *Nature Climate Change* 8 (May 21, 2018): 510–14.

21. Anthony McMichael, Alistair Woodward, and Cameron Muir, *Climate Change and the Health of Nations: Famines, Fevers, and the Fate of Populations* (New York: Oxford University Press, 2017).

22. Rok Tkavc et al., "Prospects for Fungal Bioremediation of Acidic Radioactive Waste Sites: Characterization and Genome Sequence of *Rhodotorula taiwanensis* MD 1149," *Frontiers in Microbiology* (January 8, 2018), DOI:10.3389/fmicb.2017.02528.

21. SCIENCE, IGNORANCE, AND MYSTERY

1. Hans Rosling, Ola Rosling, and Anna Rosling Ronnlund, *Factfulness: Ten Reasons We're Wrong about the World—and Why Things Are Better Than You Think* (New York: Flatiron, 2018).

2. Steven Pinker, *Enlightenment Now: The Case for Reason, Science, Humanism, and Progress* (New York: Penguin, 2018).

3. Mazhar Adli, "The CRISPR Tool Kit for Genome Editing and Beyond," *Nature Communications* 9, no. 1 (May 15, 2018): 1911, DOI:10.1038/s41467-018-04252-2.

4. Pam Belluck, "Chinese Scientist Who Says He Edited Babies' Genes Defends His Work," *New York Times*, November 28, 2018.

5. Shany Doron et al., "Systematic Discovery of Antiphage Defense Systems in the Microbial Pangenome," *Science* 359, no. 6379 (March 2, 2018), DOI:10.1126/science.aar4120.

6. John Bohannon, "A New Breed of Scientist, with Brains of Silicon," *Science*, July 5, 2017, DOI:10.1126/science.aan7046.

7. Stuart Firestein, *Ignorance: How It Drives Science* (New York: Oxford University Press, 2012).

8. Nick Lane, *The Vital Question: Energy, Evolution, and the Origins of Complex Life* (New York: W. W. Norton, 2015).

9. Antonio Damasio, *The Strange Order of Things: Life, Feeling, and the Making of Cultures* (New York: Pantheon Books, 2018).

10. Kathryn Pritchard, "Religion and Science Can Have a True Dialogue," *Nature* 537 (September 22, 2016): 451, DOI:10.1038/537451a.

FURTHER READING

Anderson, Scott C., John F. Cryan, and Ted Dinan. *The Psychobiotic Revolution: Mood, Food, and the New Science of the Gut-Brain Connection.* Washington, DC: National Geographic, 2017.

Anton, Ted. *Planet of Microbes: The Perils and Potential of Earth's Essential Life Forms.* Chicago: University of Chicago Press, 2017.

Barnosky, Anthony D. *Dodging Extinction: Power, Food, Money, and the Future of Life on Earth.* Oakland: University of California Press, 2014.

Bauerfeind, Rolf, Alexander von Graevenitz, Peter Kimmig, Hans Gerd Schiefer, Tino Schwartz, Werner Slenczka, and Horst Zahner. *Zoonoses: Infectious Diseases Transmissible from Animals to Humans.* 4th ed. Washington, DC: ASM Press, 2016.

Biddle, Wayne. *A Field Guide to Germs.* New York: Henry Holt, 1995.

Blaser, Martin J. *Missing Microbes: How the Overuse of Antibiotics Is Fueling Our Modern Plagues.* New York: Henry Holt, 2014.

Bloomberg, Michael, and Carl Pope. *Climate of Hope: How Cities, Businesses, and Citizens Can Save the Planet.* New York: St. Martin's Press, 2017.

Choffnes, Eileen R., LeighAnne Olsen, and Alison Mack, rapporteurs. *Microbial Ecology in States of Health and Disease (Workshop Summary).* Institute of Medicine. Washington, DC: National Academies Press, 2014.

Chutkan, Robynne. *The Microbiome Solution: A Radical New Way to Heal Your Body from the Inside Out.* New York: Avery, 2015.

Clark, David P. *Germs, Genes and Civilization: How Epidemics Shaped Who We Are Today.* Upper Saddle River, NJ: FT Press, 2010.

Damasio, Antonio. *The Strange Order of Things: Life, Feeling, and the Making of Cultures.* New York: Pantheon Books, 2018.

Darwin, Charles. *The Origin of Species by Means of Natural Selection of the Preservation of Favored Races in the Struggle for Life.* New York: New American Library, 2003.

Dennett, Daniel C. *From Bacteria to Bach and Back: The Evolution of Minds.* New York: W. W. Norton, 2017.

Diamond, Jared. *Guns, Germs, and Steel: The Fates of Human Societies.* New York: W. W. Norton, 1997.

Dietert, Rodney. *The Human Superorganism: How the Microbiome Is Revolutionizing the Pursuit of a Healthy Life.* New York: Dutton, 2016.

Dunn, R. *The Wild Life of Our Bodies: Predators, Parasites, and Partners That Shape Who We Are Today.* New York: Harper Perennial, 2014.

Eldredge, Niles. *Eternal Ephemera: Adaptation and the Origin of Species from the Nineteenth Century through Punctuated Equilibria and Beyond.* New York: Columbia University Press, 2015.

Firestein, Stuart. *Ignorance: How It Drives Science.* New York: Oxford University Press, 2012.

Fortey, Richard. *Life: A Natural History of the First Four Billion Years of Life on Earth.* New York: Vintage Books, 1997.

Goetz, Thomas. *The Remedy: Robert Koch, Arthur Conan Doyle, and the Quest to Cure Tuberculosis.* New York: Avery, 2014.

Hall, W., A. McDonnell, and J. O'Neill. *Superbugs: An Arms Race against Bacteria.* Cambridge, MA: Harvard University Press, 2018.

Harari, Yuval Noah. *Sapiens: A Brief History of Humankind.* New York: HarperCollins, 2015.

Holt, Jim. *Why Does the World Exist? An Existential Detective Story.* New York: Liveright, 2012.

Kolbert, Elizabeth. *The Sixth Extinction: An Unnatural History.* New York: Henry Holt, 2014.

Kolter, Roberto, and Stanley Maloy, eds. *Microbes and Evolution: The World Darwin Never Saw.* Washington, DC: ASM Press, 2012.

Lane, Nick. *The Vital Question: Energy, Evolution, and the Origins of Complex Life.* New York: W. W. Norton, 2015.

Lederberg, Joshua, Robert E. Shope, and Stanley C. Oaks Jr., eds. *Emerging Infections: Microbial Threats to Health in the United States.* Washington, DC: National Academies Press, 1992.

Mackenzie, John S., M. Jeggo, P. Daszak, and J. A. Richt, eds. *One Health: The Human–Animal–Environment Interfaces in Emerging Infectious Diseases.* Current Topics in Microbiology and Immunology 366. New York: Springer, 2013.

Mayer, Emeran. *The Mind-Gut Connection: How the Hidden Conversation within Our Bodies Impacts Our Mood, Our Choices, and Our Overall Health.* New York: HarperCollins, 2016.

McKenna, Maryn. *Big Chicken: The Incredible Story of How Antibiotics Created Modern Agriculture and Changed the Way the World Eats.* Washington, DC: National Geographic, 2017.

McMichael, Anthony J., Alistair Woodward, and Cameron Muir. *Climate Change and the Health of Nations: Famines, Fevers, and the Fate of Populations.* New York: Oxford University Press, 2017.

McNeill, William H. *Plagues and People.* New York: Doubleday, 1997.

Monosson, Emily. *Natural Defense: Enlisting Bugs and Germs to Protect Our Food and Health.* Washington, DC: Island Press, 2017.

Morris, Robert D. *The Blue Death: Disease, Disaster, and the Water We Drink.* New York: HarperCollins, 2007.

Mukherjee, Siddhartha. *The Gene: An Intimate History.* New York: Scribner, 2016.

National Academies of Sciences, Engineering, and Medicine. *Exploring Lessons Learned from a Century of Outbreaks: Readiness for 2030; Proceedings of a Workshop.* Washington, DC: National Academies Press, 2019.

Nye, Bill. *Undeniable: Evolution and the Science of Creation.* New York: St. Martin's Press, 2014.

O'Malley, Maureen A. *Philosophy of Microbiology.* New York: Cambridge University Press, 2014.

Osterholm, Michael T., and Mark Olshader. *Deadliest Enemy: Our War against Killer Germs.* New York: Little, Brown, 2017.

Otto, Shawn. *War on Science: Who's Waging It, Why It Matters, What Can We Do about It.* Minneapolis: Milkweed Editions, 2016.

Pennington, Hugh. *Have Bacteria Won?* Cambridge: Polity Press, 2016.

Pinker, Steven. *Enlightenment Now: The Case for Reason, Science, and Humanism.* New York: Penguin, 2018.

Piot, Peter. *AIDS Between Science and Politics.* New York: Columbia University Press, 2015.

Quammen, David. *Spillover: Animal Infections and the Next Human Pandemic.* New York: W. W. Norton, 2012.

————. *The Tangled Tree: A Radical New History of Life.* New York: Simon & Schuster, 2018.

Rhodes, John. *The End of Plagues: The Global Battle against Infectious Disease.* New York: Palgrave Macmillan, 2013.

Rosling, Hans, Ola Rosling, and Anna Rosling Rönnlund. *Factfulness: Ten Reasons We're Wrong about the World—and Why Things Are Better Than You Think.* New York: Flatiron, 2018.

Shah, Sonia. *Pandemic Trafficking Contagions, from Cholera to Ebola.* New York: Sarah Crichton Books, 2016.

Smolinski, Mark S., Margaret A. Hamburg, and Joshua Lederberg. *Microbial Threats to Health: Emergence, Detection, and Response.* Washington, DC: National Academies Press, 2003.

Thomas, Chris D. *Inheritors of the Earth: How Nature Is Thriving in an Age of Extinction.* New York: PublicAffairs, 2017.

Verstock, Frank T., Jr. *The Genius Within: Discovering the Intelligence of Every Living Thing.* New York: Harcourt, 2002.

Wilson, Edward O. *Biophilia.* Cambridge, MA: Harvard University Press, 1984.

————. *Genesis: The Deep Origin of Societies.* New York: Liveright, 2019.

————. *The Meaning of Human Existence.* New York: Liveright, 2014.

Yong, Ed. *I Contain Multitudes: The Microbes within Us and a Grander View of Life.* New York: HarperCollins, 2016.

Zimmer, Carl. *A Planet of Viruses.* Chicago: University of Chicago Press, 2011.

Zinsser, Hans. *Rats, Lice, and History.* Boston: Little, Brown, 1934.

INDEX